Risk Assessment

1. Haz Identification
2. " Characterization.
3. Exposure Assessment
4. Risk Characterization.

MW01630405

FSM M
20 - Measuring & Monitoring
21 - Site security to control operational conditions.
22 - Product label
23 - Trace.
24. - Pack Material.
18 - complaint.
19 - Incident.
16 - Supplier Performance & approval

Assessments of effectiveness of CA plan include:

Determine controls are adequate for a capable process

Controllable hazards — HACCP seeks to prevent, eliminate or Reduce to an acceptable level — hazards

Checklist — help guide auditor through audit Perform

Swot analysis — tool for keeping management apprised of emerging international ~~Standards~~ hazards as part of revi

Consumer use info — identify vulnerable groups

6 Step audit interview proc

1. put at ease.
2. Explain purpose.
3. find out what they do.
4 Analyze ".
5 tentative conclusion
6. Explain next step.

3 levels of qualification to provide document evidence of compliance

1. Immediate org level (local)
2. professional.
3. Conformity assessment.

CCP decision tree = identify CCP + Separate them from non-Safety related controls.

Organization effective = how effective in achieving Results that u intend to produce

System & process effe = How a System or Process can meet the demand under Specified conditions

Conformance to Req = when evaluating raw data to specs, one is evaluating against a conformance to requirements.

Performance measure = Systemic way to track, manage, report progress towards the attainment of goals & objectives

HACCP System must be verified.

HACCP Plan validated (Befor implement)

HACCP System = Result of implementation of plan.

The ASQ Certified Food Safety and Quality Auditor Handbook

purpose for audit of decisions = Compare inspector results with worker

Fourth Edition

Steven Wilson, Editor

ASQExcellence
Milwaukee, Wisconsin

Published by ASQExcellence, Milwaukee, WI
Produced and distributed by Quality Press, ASQ, Milwaukee, WI

Publisher's Cataloging-in-Publication data

Names: Wilson, Steven, 1959-, editor.
Title: The ASQ certified food safety and quality auditor handbook, fourth edition / Steven Wilson, Editor.
Description: Includes bibliographical references and index. | Milwaukee, WI: ASQ Excellence, 2021.
Identifiers: LCCN: 2021933376 | ISBN 978-1-951058-18-0 (hardcover) | 978-1-952236-12-9 (hardcover) | 978-1-951058-19-7 (epub) | 978-1-952236-13-6 (epub) | 978-1-951058-20-3 (PDF) | 978-1-952236-14-3 (PDF)
Subjects: LCSH Food adulteration and inspection—Handbooks, manuals, etc. | Food handling—Safety measures—Handbooks, manuals, etc. | Food industry and trade—Safety measures—Handbooks, manuals, etc. | BISAC TECHNOLOGY & ENGINEERING / Food Science / Food Safety & Security
Classification: LCC TX531.A87 2021 | DDC 363.19/2—dc23

ASQ and ASQExcellence advance individual, organizational, and community excellence worldwide through learning, quality improvement, and knowledge exchange.

Attention bookstores, wholesalers, schools, and corporations: Quality Press and ASQExcellence books are available at quantity discounts with bulk purchases for business, trade, or educational uses. For information, please contact Quality Press at 800-248-1946 or books@asq.org.

To place orders or browse the selection of ASQExcellence and Quality Press titles, visit our website at: http://www.asq.org/quality-press

Printed in Canada.

25 24 23 22 21 FC 5 4 3 2 1

Quality Press
600 N. Plankinton Ave.
Milwaukee, WI 53203-2914
Email: books@asq.org
Excellence Through Quality™

Table of Contents

List of Figures and Tables

Foreword

Management of food safety and quality has evolved greatly over the years, particularly since the implementation of Hazard Analysis Critical Control Point (HACCP). The certifications offered by the American Society for Quality (ASQ) and the American Society for Quality Excellence (ASQE) have worked to keep up with this evolution. As noted in the last edition of the *The Certified HACCP Auditor Handbook*, this certification first started as an add-on to the Certified Quality Auditor and soon became a stand-alone certification. Our next evolution sees the certification going beyond HACCP and toward a more comprehensive thought process in systems management. With this evolution, the auditor of such systems must also evolve. The body of knowledge was again updated to encompass the foundations developed in earlier certifications and also to direct the auditor toward the changes expected by industry and the public. In 2004, the ASQ Certification Board approved the Certified HACCP Auditor as a stand-alone certification. It is important for the auditor not to be a holder of the technical knowledge of these various systems, for they are becoming too complex, but to utilize critical thinking to arrive at a decision as to how the system is operating against the various standards that are being applied.

The Food, Drug, and Cosmetic Division of ASQ is committed to increasing customer satisfaction and continuous improvement by identifying, communicating, and promoting knowledge and the use of management concepts, technologies, and regulations as they relate to quality principles in all functional areas of the food, drug, and cosmetic industries. Their efforts are designed to ensure that quality and safety will be sustained for future generations.

Steven Wilson
Project Lead
Washington, DC

Acknowledgments

Paul K. Mignon is acknowledged as a contributing author of the Animal Food chapter. Additionally, the following individuals are recognized for their technical contribution, review, and hard work getting this handbook to completion: Tatiana Miranda, Jen Stephenson, Aura Stewart, Rosemarie Christopher, and Lisa El-Shall. A final acknowledgment to John Surak, who started it all.

Part I

An Introduction to HACCP

Chapter 1

History and Overview of HACCP: Primitive and Modern Food Preservation Methods

Humans have been concerned with the availability of food from the dawn of their existence. Prehistoric humans were hunters and gatherers who needed to find and catch food. As time passed, humans began to grow and preserve their own food. And still later in history, humans became concerned with preparing, conserving, and maintaining a steady supply of food. Advances in social organization from small- to large-group living and from nomadic hunting and gathering to communal life in a fixed place paralleled the need for a reliable source of food.

The methods for saving food for the proverbial rainy day may have begun with air drying, salting, and the use of spices and herbs, and then advanced to more sophisticated technologies such as canning and freezing. Advances in technology took thousands of years, and many occurred by accident. Transfer of technology occurred slowly because of a lack of communication and commerce among geographically dispersed societal groups.

However, as trade and communication increased, primitive food preservation technologies were transferred from one culture to another. Two examples of this are: (1) the drying of grain and the storage of that grain in large granaries in the Middle East and Africa; and (2) the bringing of pasta, a different form of preserved grain, from China to Europe by the explorer Marco Polo. As commerce became more important, laws were developed to control the quantity and quality of traded goods, including food, as well as services. The first comprehensive written code was set down by Hammurabi, 1772 BCE. Later, laws were set forth in the Torah and the Holy Bible (see Leviticus, Chapter 11 and Deuteronomy, Chapter 7). Since it is impossible to chronicle all the advances in food preservation technology here, only some of the more important advances that have influenced modern techniques for making safe food available to all are discussed. Not all modern methods are new: primitive preservation methods such as drying, salting, and smoking are still used. Other currently used methods for preserving foods include heat preservation by canning in hermetically sealed containers, pasteurization, freezing, freeze-drying, and air drying. The use of these preservation methods was the first documented preventive response to anticipated hazards. The anticipatory and preventive methodology that became known as HACCP began here.

HACCP PREDECESSORS

At some point, scientists discovered that microscopic organisms could cause food spoilage. This led to the theory that food could be preserved if the spoilage organisms could be destroyed and kept from reentering the food product. For this to occur, the temperature and water content of food had to be reduced to levels that would not support the growth of spoilage organisms. Louis Pasteur and Clarence Birdseye were leaders in food processing technology. Additionally, research in industrial areas unrelated to the production of food by quality gurus such as Walter A. Shewhart, Joseph M. Juran, and W. Edwards Deming was adapted by others and applied to the control of quality in the production and preservation of food.

One of the earliest collaborative efforts of industry and government addressed the problem of milk-borne disease. In the 1920s, two industry associations and one professional association developed uniform standards for fittings used in dairy and food handling equipment. The standards for fittings became known as 3-A Standards. "Since 1944, the 3-A Program has included representation from suppliers and equipment fabricators, all national dairy processing associations, the U.S. Department of Agriculture (USDA), the U.S. Public Health Service (USPHS), and state regulatory agencies."[1]

Milk safety was accomplished by controlling the following factors, which are elements of what is known today as the Hazard Analysis Critical Control Point (HACCP)[2] approach to product safety:

- The health and sanitation of the dairy herd
- The times of collection and temperatures of milk from collection to processing
- The use of a terminal heat treatment to reduce microbial content
- The standardization of equipment
- The scrupulous cleaning of the processing plant and equipment
- The temperature of the processed product after pasteurization and while in transit and in storage

As the populace of the United States shifted from agrarian to urban living, there was an increasing need to process foods for mass transport and consumption in cities. The earlier emphasis on raw agricultural products shifted to processed products. Currently, food is prepared outside the home for consumption in homes, restaurants, schools, hospitals, nursing homes, and prisons; aboard airplanes, ships, and trains; during camping or wartime; and even in space vehicles.

These new modes of consumption required the development of new methods for use in the preparation, packaging, and storage of foods to ensure the availability of food that is safe, nutritious, and wholesome.

HACCP AND THE SPACE PROGRAM

In the late 1950s, the National Aeronautics and Space Administration (NASA) saw the need for special foods for space travel.[3] The early space vehicles were small, and there was room for neither standard kitchen appliances—refrigerator, stove, freezer—nor the pantry, cupboards, and countertops commonly used for the storage and preparation of foods. In addition, concerns existed about the kinds of food that an astronaut could take on a space journey that would provide proper nutritional, gustatory, and safety properties. It was also important that the space vehicle and its contents not introduce harmful microorganisms into space.

Before the dawn of the space age, food quality and safety were controlled mainly by finished product inspection. But NASA wanted assurances that safety was built into the design of the food manufacturing process. In the early 1960s, the Pillsbury Company was asked to develop the first space foods, as well as to design a system for controlling the safety of space foods, used first for the *Mercury* flights and later for the *Gemini* and *Apollo* flights. NASA also was concerned about food crumbs floating in the cabin and fouling the instruments of the space vehicles. Pillsbury easily solved the crumb problem by coating bite-sized pieces of food to prevent crumbling. But they had a more daunting task in ensuring the bacterial quality of space foods.

To ensure that foods used in the space program were safe, Pillsbury developed the HACCP system. HACCP was designed to prevent safety hazards. By systematically evaluating the ingredients, environs, and processes used to fabricate a food, identifying areas of potential risk, and determining the critical control points (that is, those points in the process that must be controlled to prevent an unacceptable risk), the manufacturer would have the assurance of process and product integrity.

As the NASA flights became longer, additional logistical requirements challenged Pillsbury to refine the HACCP system. Pillsbury collaborated with NASA and the U.S. Army's Natick Laboratories to develop HACCP as a proactive system for manufacturing and supplying safe foods for space travelers. By the time the Eagle landed and man set foot on the moon in 1969, Pillsbury had developed HACCP as we know it today (see Figure 1.1).

In 1967, the U.S. Food and Drug Administration (FDA) and the food industry began a pilot self-certification program that was designed to incorporate HACCP concepts into the food manufacturing process. Participants in the pilot program were required to share information about their products, processes, and quality control, including planned changes, with the FDA. The overall objectives were: (1) to have the industry participants exercise more control over their operations; and (2) to give the FDA a better view of the controls exercised by the industry participants than a random inspection would allow. This program was ahead of its time. It was not politically correct then, so it felt the wrath of Congress and consumers, neither of whom believed that industry was capable of "self-certifying." The FDA altered the program and eliminated the name "self-certification," calling it instead the "cooperative quality assurance program." However, the revamped program (later discontinued) retained HACCP at its core.

HACCP involves seven principles:

1. *Analyze hazards.* Potential hazards associated with a food and measures to control those hazards are identified. The hazards could be biological, such as a microbe, chemical, such as a toxin, or physical, such as ground glass or metal fragments.
2. *Identify critical control points.* These are points in a food's production—from its raw state through processing and shipping to consumption by the consumer—at which the potential hazard can be controlled or eliminated. Examples are cooking, cooling, packaging, and metal detection.
3. *Establish preventive measures with critical limits for each control point.* For a cooked food, for example, this might include setting the minimum cooking temperature and time required to ensure the elimination of any harmful microbes.
4. *Establish procedures to monitor the critical control points.* Such procedures might include determining how and by whom cooking time and temperature should be monitored.
5. *Establish corrective action to be taken when monitoring shows that a critical limit has not been met.* An example would be reprocessing or disposing of food if the minimum cooking temperature is not met.
6. *Establish procedures to verify that the system is working properly.* An example would be testing time-and-temperature recording devices to verify that a cooking unit is working properly.
7. *Establish effective record keeping to document the HACCP system.* This includes records of hazards and their control methods, the monitoring of safety requirements, and action taken to correct potential problems. Each of these principles must be backed by sound scientific knowledge; for example, published microbiological studies on time and temperature factors for controlling foodborne pathogens.

Figure 1.1 What is HACCP?

Source: U.S. Food and Drug Administration, "A State-of-the-Art Approach to Food Safety," *FDA Backgrounder* (August 1999).

APPLICATION OF HACCP TO OTHER INDUSTRIES

In the early 1970s, Pillsbury transferred the HACCP concept from the space program to production in its commercial food plants. This technology also was transferred to the FDA in a contract for training FDA personnel in HACCP methodology. In the early 1960s, the State of California's Department of Health Services pioneered the application of HACCP principles in its canning industry. It became the prototype for a regulation—Code of Federal Regulations (CFR) title 21, part 113—promulgated by the FDA in the mid-1970s in response to an industry petition. This regulation incorporates HACCP methodology to govern the production of low-acid canned foods in hermetically sealed containers. HACCP is now mandatory in the FDA program for food safety for fish and shellfish (21 CFR part 123) and other products.

What made HACCP so popular after it languished for so long? After all, HACCP had been used in food processing plants since the late 1960s but had not been adopted on a large scale. Perhaps the climate was right—public health officials were concerned about emerging pathogens, and consumers and industry were concerned about food safety. These sectors with converging interests knew there had to be a better way to ensure the safety of foods. Similarly, the economy had become globalized and food safety had become an international, rather than simply a national, concern.

A succession of reports by three prestigious groups opened the door to HACCP on a global basis:

- The National Academy of Sciences report, *Microbiological Criteria for Foods and Food Ingredients,* 1985
- Report of the International Commission for the Microbiological Specifications for Food (ICMSF), 1988
- The Codex Alimentarius Commission's *Guidelines for the Application of the Hazard Analysis and Critical Control Point (HACCP) System,* 1991, adopted by the Joint FAO/WHO Codex Commission, 20th Session, 1993

Today, technology from the space meal systems has been transferred to the private sector and is being used in meal systems for the elderly. The use of HACCP has been truly merged with the requirements of a management system. As a result, a number of food safety management standards have been published, including the following:

- ISO 22000 and the ISO 22002 series of standards
- BRC Global Standard for Food Safety
- IFS Food
- SQF Code (Food Industry)

The Global Food Safety Initiative (GFSI) has developed a benchmarking process to determine if an audit scheme meets the ever-evolving science-based food safety requirements that are deemed necessary for the production of safe food. The benchmarking scheme consists of two parts: (1) the food safety requirements; and (2) the requirements that accreditation bodies and certification bodies must follow to provide certification. It should be noted that ISO 22000 is not a certification scheme. However, FSSC 22000 is a certification scheme that uses ISO 22000 and the ISO 22002 family of standards. FSSC 22000 has been benchmarked as meeting all the GFSI requirements.

The Food Safety Modernization Act (FSMA) was signed into law on January 4, 2011. This law required the FDA to develop regulations requiring that all non-meat and non-poultry establishments implement hazard analysis and risk-based preventive controls. In 2004, the European Union (EU) developed regulations that require all food processors in the EU to implement HACCP.

Currently, the seven principles of HACCP are being applied on a pilot scale in the medical device industry to increase the safety of its products.

MOVING TO A COMPREHENSIVE FOOD SAFETY AND MANAGEMENT SYSTEM

Food safety and quality systems were not always preventive in nature. Inspection and testing in the final product form were the cornerstone of any good quality program and ultimately a good food safety program. Over time, quality assurance systems evolved from testing for defects to preventing defects from occurring in the first place. Further, it was the necessity of the NASA space program, discussed previously, and the development of HACCP that moved food safety to a preventive structure.

Now the focus on food safety and quality is preventive in nature. The previous discussion on prerequisite programs demonstrates all that goes into preventing food safety concerns where possible. Only then does a HACCP system get implemented and begin to thrive. The first HACCP systems were specific to a need: microbiological for NASA, then as a means for demonstrating control in exports of goods. The seafood HACCP regulations were the first major push by the U.S. FDA for a food product, with juice HACCP coming soon after.

With the FSMA in 2011, the FDA required implementation of preventive controls for all food products. These regulations required food production firms to utilize the same methods of HACCP in a general way for the prevention of food safety issues.

Under these regulations, firms are required to implement:

- Hazard analysis: Identify any known or reasonably foreseeable hazards and determine if those hazards require a preventive control.
- Preventive controls: If an identified hazard requires a preventive control, develop and implement a control to significantly minimize or prevent the hazard. These controls must include process controls, food allergen controls, sanitation controls, and other controls not previously identified that are deemed necessary to prevent or minimize the hazard.

Once the firm has identified these controls, it must perform monitoring and corrective action to maintain its control procedures. Firms are required to implement specific training programs to keep their personnel properly qualified. Supply chain programs are also required, as well as documentation of the controls put in place. In addition, firms must have in place a written recall plan, including periodic effectiveness checks.

These requirements mimic the prerequisite programs and HACCP systems described by the National Advisory Committee on Microbiological Criteria for Foods (NACMCF) and Codex, as well as most systems implemented by governments across the globe.

NOTES

1. More information about the 3-A Program is available online at http://www.3-a.org.
2. The terms "hazard analysis critical control point system" and "hazards analysis critical control point concept" are used interchangeably.
3. More information about the U.S. space program is available online at http://www.spaceflight.nasa.gov/history.

Chapter 2

Prerequisite Programs

A solid foundation of prerequisite programs (PRPs) is necessary to support a HACCP and food safety management system. Many prerequisites are broader in scope than product safety; some are directly linked to HACCP, while others are managed outside the HACCP system.

Controls for many hazards, excluding critical control points (CCPs), reside in PRPs. While the relatively few CCPs in a process should be addressed in the HACCP summary table, the importance of the broader prerequisite controls must not be overlooked. PRPs must first be systematically identified based on risk evaluation of the environment and operating conditions and then examined as part of the overall HACCP system implementation. An early step in implementing HACCP is monitoring mandated prerequisites in the organization's regulatory environment.

In preparing to develop or audit a HACCP plan, it is necessary to assemble a list of all pertinent PRPs (including any mandated by regulation) and to locate and review related procedures, work instructions, and records. Many, if not all, of these will reside in the company's current quality, food safety, and operating manuals. Likewise, in auditing it is essential for audit teams to reference and familiarize themselves with regulations where pertinent. These regulations pose an appropriate standard upon which to judge compliance and effectiveness.

Each industry segment under consideration must implement the PRPs necessary to assure a safe operating environment. PRPs create the environment for the safe production of a food. In some industry segments, pertinent regulations prescribe certain PRPs, thus effectively linking them with HACCP.

A company may elect to utilize existing system or process auditing teams to evaluate PRPs in conjunction with HACCP implementation. Common PRPs often are listed along with generally stated "standards" for the audit team to use as a reference in compiling audit checklists. Those listed below identify the foundations for consideration when planning or auditing a food safety and HACCP system and should be considered when compiling an audit checklist.

- *Plant and grounds construction.* Flow and traffic control should be implemented to reduce the potential for cross-contamination.
- *Processing equipment design.* The equipment should be constructed according to sanitary design principles and should be properly maintained to prevent deterioration and product contamination.

- *Pre-start-up inspection.* Cleaning, sanitizing, and housekeeping efficacy should be verified prior to the start of production.
- *Specification system.* The company should maintain written specifications for all ingredients, package materials, and finished products.
- *Vendor/material control.* Effectiveness of vendors' quality and food safety systems should be verified.
- *Master cleaning procedures and sanitization.* Cleaning and sanitizing activities related to product safety should be scheduled and monitored.
- *Storage, shipping, and handling.* Supplies/materials/finished products must be stored, shipped, and handled properly. This includes ensuring that temperature is controlled as appropriate and that cross-contamination is prevented throughout distribution.
- *Chemical control.* Potentially toxic, caustic, or injurious substances whose use is unavoidable in and around product manufacturing must be effectively segregated.
- *Recall and material traceability.* The ability to identify or recover stock at any stage of production and distribution must be ensured through appropriate coding and records of use.
- *Pest control.* An appropriately qualified and licensed pest control operator must conduct an effective program in concert with applicable regulations.
- *Operator/employee training.* Operators must be appropriately trained to properly conduct their assigned tasks. In addition, the company must ensure that the operators have the competencies to conduct their assigned tasks.
- *Personnel hygiene and employee facilities.* Operators, management, and visitors must conduct themselves in a manner that prevents the contamination of food products. In addition, employee facilities must be maintained in a manner that prevents cross-contamination.
- *Good laboratory practices (GLPs).* Operators using laboratory testing in their process must abide by principles to assure the integrity and quality of the testing, including a testing continuity plan.
- *Calibration and standardization.* Calibration schedules should be established for all CPs and CCPs in the manufacturing process, as well as for laboratory and analytical equipment used in monitoring or testing.
- *Environmental monitoring.* Microbiological evaluation of the processing and production "environment" (drains, air-handling systems, and so on) should be conducted as appropriate to control pathogens and spoilage organisms, keeping food safety and quality in mind.

- *Allergen controls.* In food and related industries, cross-contamination of products with undeclared allergens (substances known to induce an adverse immunological response in susceptible individuals) should be minimized through labeling control and prevention of physical cross-contamination.
- *Foreign material control.* In various industry sectors, screens, magnets, metal detectors, and other foreign material exclusion devices should be employed.
- *Waste disposal.* The company must make appropriate provisions for the proper segregation, storage, and removal of waste.
- *Utilities.* The company must provide appropriate utilities (water, boiler chemicals, air quality and ventilation, compressed air or gases, lighting, electricity or other sources of energy, and so on) to minimize the risk of product contamination and cross-contamination.
- *Rework.* Rework should be stored, handled, and used in such a way that product safety, quality, and traceability and regulatory compliance are maintained.
- *Cross-contamination.* Programs must be in place to prevent, control, and detect physical, chemical, and microbial contamination.
- *Crisis management.*
- *Good agricultural practices (GAPs).* Programs that include any fruits or vegetables must ensure they are produced, packed, handled, and stored as safely as possible to minimize risks of microbial food safety hazards.

HACCP implementation involves a close examination of PRPs rather than a cursory review. PRPs are an essential part of the HACCP system. They function to reduce the complexity of the HACCP plan. Thus, they need to be continually verified to ensure their effectiveness.

The implementation of a HACCP system will probably result in the need to upgrade certain PRPs. The act of doing a thorough hazard analysis (HACCP principle #1) often brings new knowledge or different perspectives about hazards to the company. This analysis often appropriately concludes that control of some hazards must revert to PRPs. These programs, which may have been previously installed in the absence of a HACCP perspective, may need to be revised.

EVOLUTION OF PREREQUISITE PROGRAMS

Many government regulations and food industry guidelines regarding food safety have been established around PRPs or procedures that address operational conditions. PRPs provide the foundation for a HACCP system. They also simplify the HACCP plan by reducing the number of CCPs. This chapter discusses the evolution of, and identifies common categories for, food safety PRPs.

PRPs originated from food regulations and voluntary food industry programs. These programs are also known as the current good manufacturing practices, or GMPs. For FDA-regulated companies, the GMPs are codified in CFR title 21, part 117. For U.S. Department of Agriculture (USDA) Food Safety and Inspection Service (FSIS) regulated companies, the GMPs are codified in CFR title 9, part 416.

GMPs were established to help define for the food industry the minimum sanitary conditions for processing safe food products. They include such areas as personal hygiene, operational practices, cleaning and sanitation, water safety, foreign material control, and sanitary design. The GMP programs in most quality-oriented firms and the GFSI food safety schemes exceed the federal regulatory requirements.

The seafood HACCP regulation, CFR title 21, part 123, requires GMPs and sanitation standard operating procedures (SSOPs) as prerequisite requirements for the HACCP program.[1] These address eight areas: pest control; employee hygiene; water safety; protection from adulterants; prevention of cross-contamination; condition of hand washing, hand sanitizing and toilet facilities; condition and cleanliness of food contact surfaces; and labeling, storage, and use of toxic compounds. USDA-FSIS also has established its version of SSOPs for the meat and poultry industries.[2] These prerequisites are divided into two categories: preoperational procedures and operational sanitation. The preoperational procedures include the cleaning of food contact surfaces, cleaning of equipment, and cleaning of utensils, while the operational SSOPs include equipment cleaning, employee hygiene, and proper product handling. In 1998, the FDA published the *Guide to Minimize Microbial Food Safety Hazards for Fresh Fruits and Vegetables.*[3] This publication addresses prerequisites or GAPs used to minimize biological hazards common to the growing, harvesting, washing, sorting, packing, and transporting of fresh fruits and vegetables sold in an unprocessed or minimally processed state.

Numerous industry guidelines that specify higher standards than those imposed by regulatory GMPs have evolved over time. Originally developed for the purpose of self-inspection and continual improvement, these standards are now also used by many companies for the purpose of supplier certification. A sample guideline is AIB International's *AIB Consolidated Standards for Inspection of Prerequisite and Food Safety Programs.*[4] This document has been further developed to address numerous sectors of the food industry, such as dairy products, fresh-cut produce, raw (unprocessed) fruits and vegetables, and food distribution centers. Using a sector-specific approach, these guidelines develop and emphasize food safety issues for the specific food industry and the PRPs most applicable to food safety.

As previously noted, food safety PRPs can be described and categorized in many ways, depending on the regulatory perspective and industry sector. CCPs—not PRPs—normally are used to address significant food hazards, but even this varies from one industry sector to another. In one industry segment a certain prerequisite may be of minor importance, while in another the same prerequisite may be essential to ensuring product safety.

If the PRP is used to create an environment where a hazard is not likely to occur, it can be essential to ensuring product safety. When this occurs, the PRP must be properly designed, implemented, and maintained. In addition, the company should take actions to check, verify, and document its continued effectiveness.

TYPES OF PREREQUISITE PROGRAMS

Because of the various perspectives described previously, many different categories have been developed for food safety PRPs. From a regulatory HACCP perspective, six key prerequisites should be implemented: GMPs, trace and recall, cleaning and sanitation, pest control, chemical control, and customer complaints—food safety.[5] Sperber et al. describe eight PRP categories: facilities, raw material controls, sanitation, training, production equipment, production controls, storage and distribution, and product controls.[6] The NACMCF 1997 HACCP guideline lists 11 PRPs: facilities; supplier control; specifications; production equipment; cleaning and sanitation; personal hygiene; training; chemical control; receiving, storage, and shipping; traceability and recall; and pest control.[7]

In an effort to consolidate these different programs while still addressing the diverse needs of the food industry, this book will describe PRPs according to the following categories: chemical control; cleaning and sanitation; microbiological control; sanitary design and engineering; preventive maintenance; trace and recall; pest control; receiving, storage, and shipping controls; supplier control; water supply; air and gas supply; food safety training; equipment calibration; customer complaints—food safety; and audits and inspection programs. This list is comprehensive but not exhaustive, and many of the categories overlap in the web of quality systems used to manage food safety.

Personal Hygiene

Personal hygiene programs focus on preventing product contamination caused by the interface of employees and food process zones. Companies typically establish GMP rules and train employees in compliance. Clothing and garment rules cover the use of clean, appropriate outer garments (limitations on buttons, fuzzy sweaters, and so on) or uniforms, the use of hair and beard restraints, restrictions on hair ties and pins, limitations on the wearing of jewelry, the use of proper footwear, the control of items in top pockets (or elimination of pockets), and the use of strong perfumes, false eyelashes, false fingernails, and fingernail polish. A trend that could result in potential food hazards and should be addressed through these policies is exposed body-piercing jewelry.

Personal hygiene also addresses disease control. Without proper safeguards, biological hazards such as *Shigella* and hepatitis A may be transmitted from personnel to food products. Employees must wash and, in most cases, sanitize their hands before reporting to their workstations, after use of restrooms, after breaks and lunch, and after coughing, sneezing, or touching the face. For sensitive product areas, some companies require the use of protective devices such as gloves, sleeve guards, and face masks. No person with boils, sores, open and infected wounds, or a food-transmittable disease should be allowed to work in food process areas; illness should be reported to supervisory personnel. Supervisory personnel should be trained to recognize signs of biological hazards, such as a jaundiced appearance in individuals with hepatitis A. It is also important to distinguish between contagious illnesses that could be hazardous to coworkers and infectious agents that are a threat to food products.

GMPs also address eating, drinking, gum and tobacco chewing, smoking, and using toothpicks near food processing areas or "product zones." Food operations typically permit drinking water only from fountains/coolers or plastic water bottles in these sensitive areas, while providing break areas, smoking areas, and lunchrooms for other activities.

Best practices and new technologies in personal hygiene include infrared activated toilets and sinks, automated hand sanitizers, automated floor foamer (a spray sanitizer replaces boot dips and forklift wheel dip), and posted hand wash signs in restrooms, in addition to incorporating management of employee practices through behavioral management theory.

Good Operational Practices

Procedures that prevent contamination of food during handling and transfer of products fall into the good operational practices category. This includes keeping all ingredients and finished products off the floor, eliminating and/or cleaning up spills and leaks, cleaning ingredient containers before use, handling rubbish and food waste properly, discarding any food that falls on the floor, and using good housekeeping in process areas.

Good housekeeping includes the proper storage of parts, equipment, and utensils after use; storing personal effects and clothing in lockers or other designated areas; hanging air and water hoses back on designated racks and reels after use; and other activities that prevent intrusion of potential contaminants into process areas. The work area should be maintained in a reasonably sanitary condition, with a minimum of operational debris. Dedicated scoops and other utensils should be used for specific raw materials and ingredients. Ingredients and process aids should be properly labeled and stored after use to prevent cross-contamination.

Foreign Material Control and Glass Control

Procedures and equipment should be utilized to prevent the inclusion of objectionable or harmful foreign objects into food products. Objectionable items include burned product, hair, insects, and paper, while harmful contaminants include metal, glass, hard plastic, and wood splinters. Harmful contaminants may cause traumatic injury such as laceration of the mouth, tongue, throat, stomach, or intestine, as well as injury to the gums and teeth. Measures should be implemented to prevent the inclusion of metal or other extraneous material in foods. According to FDA policy, food is considered adulterated if it is ready to eat and contaminated with hard or sharp objects that measure 7–25 mm in length.[8] Customer and market-driven requirements in this area are more demanding and usually require preventing inclusion of metal contaminants bigger than a sphere of 1–3 mm in diameter.

Foreign materials may originate from raw materials and ingredients, food processing equipment, the food plant environment (walls, ceiling, and so on), employees, food service product preparation, or retail tampering. Since most food processing equipment is fabricated from metal materials, metal is a universal physical hazard in most segments of the food industry.

Equipment utilized to control metal hazards includes magnets, filters, traps, and electronic metal detectors. Nonmetallic foreign objects such as stones, bones, wood, glass, and insects can be controlled through the use of sifters, product grading screens, rock traps, de-stoners, wash tanks, aspirators, and x-ray detection equipment. In cases where equipment may not be sensitive or sophisticated enough to remove foreign objects, inspection conveyors may be required so employees can manually remove contaminants. Regardless of the equipment or procedure used to detect and remove foreign materials, findings should be logged on an ongoing basis and sources of contamination should be investigated. There is debate as to whether this is a prerequisite program, a preventive control, or should be considered as a CCP. Often it is all three, and a comprehensive metal control program is suggested.

Standard operating procedures also should be used to prevent physical hazards. Food plants should adopt a glass policy that outlines requirements for shielding fluorescent tubes and light bulbs in process areas; for protection or removal of glass gages, emergency lights, thermometers, and wall clocks; for control of glass containers in process areas; for the safe use of laboratory glassware; and for handling breakage of glass packaging materials.

Self-inspection also is used to identify and control physical hazards. Regardless of whether equipment or procedures are used, there always must be a method to evaluate physical contaminants. A process should be in place to determine if contaminants are incidental or continuous in nature and if further corrective actions should be taken.

Best practices for foreign material control include computerized networked metal detector management equipment, shielded mercury vapor bulbs, employee health screening, glass breakage programs for glass pack lines, electronic inspection equipment for glass bottles and jars (100% glass breakage inspection), and washing of incoming glass jars and bottles.

CHEMICAL CONTROL

Chemical hazards include such items as sulfites, yellow #5 (tartrazine), sanitation chemicals, allergens, mycotoxins, pesticides, refrigerants, solvents, acid, caustics, and sanitizers. They may originate from raw materials such as histamines in scromboid fish, from allergenic ingredients such as peanuts, from food additives such as sulfites, and from chemicals used in food plants such as ammonia (a refrigerant) and sodium hypochlorite. In general, chemical control relates to procedures used for the receipt, storage, use, disposal, and record keeping of chemicals needed for processing, sanitation, pest control, and maintenance. Due to the broad array of potential hazards, preventive PRPs identified here include sanitation chemical handling, process aid chemical control, plant pesticide control, maintenance chemical control, agricultural chemical control, and allergen control.

Sanitation Chemical Handling

The first step in any food plant chemical control program must be an inventory or register of all approved chemicals. This master list should include sanitation chemicals, maintenance chemicals, process aids, chemical ingredients, and pesticides used at the plant. Food plants typically maintain this master list as part of an OSHA-required hazard communication program.

Cleaning chemicals and sanitizers should be appropriate for use in food processing areas. In the past, many companies relied on the USDA screening and rating system for chemical products approved for use in meat and poultry facilities, but this program has been discontinued. Chemicals not previously rated by the USDA as safe for food plants will require labeling for food plants or supplier or third-party certification as proof that they are food grade or safe for a food plant.

Primary containers of chemicals should be stored in a segregated, locked, and identified area, rather than near food process zones. Some chemicals, such as acids and chlorine compounds, may require physical segregation. Storage areas should have spill containment equipment, such as containment walls or spill pallets, in the event of container breakage. All containers should be closed and secured after dispensing. Secondary containers in regular use near food zones should be stored a safe distance from the food zone and closed after each use to guard against spillage.

In facilities that utilize clean-in-place equipment, piping used for chemicals should be identified with signage or labels. These pipes should not run directly above exposed product zones. Chemical makeup tanks must also be appropriately identified.

The contents of all primary and secondary chemical containers and application equipment must be properly identified. Identification includes the name of the chemical and the hazard rating, from the material safety data sheet (SDS). Before use, cleaning chemicals and sanitizers generally require dilution to proper concentrations. Dilution rates usually are specified on product labels. Test kits are available to test concentrations of many of these chemicals, such as chlorine, caustics, and quaternary ammonium compounds. Where equipment is used to automatically dilute chemicals, calibration of the equipment should be a regular practice. Biocides are categorized as pesticides by the U.S. Environmental Protection Agency (EPA). They are regulated by federal pesticide law and carry an EPA registration number. Use of these materials should be regularly logged to ensure compliance with the labeled application rate. Finally, the application of these chemicals must be carefully controlled. Applications must not be made to food products or to process zones during production. Written procedures should be developed to identify proper dilution, testing, and application for all sanitation chemicals.

Best practices and new technologies in sanitation chemical handling include the use of color-coded buckets and other containers for the use of different types of chemical solutions, automatic chemical dilution and dispensing units, waterproof labels on containers in wet areas, spill control kits, spill control pallets, chemical storage areas with spill containment barriers, and employee safety programs (hazard communication, lockout/tag-out, process safety management, and so on).

Process Aid Chemical Control

During the processing of foods, process aids such as fungicides, microbicides, disinfectants, defoamers, and chelating agents are used. These include items such as hypochlorites, chlorine gas, ozone, mineral oil, peracetic acid, sprout inhibitors, sulfites, and chlorine. All of these chemicals should be food grade. Any non-food-grade chemicals used should face regulatory scrutiny and sampling. If contaminants are found, regulatory actions might follow. Labels and safety data sheets (SDSs) should be on file for each chemical. As with sanitation chemicals, storage should be controlled, and all materials should be properly labeled.

Application of processing chemicals should be closely monitored. Those chemicals that carry an EPA registration label must be applied only at the rate listed on the label, and applications should be logged to indicate amount used, where used, rate of application, concentration, date, name of applicator, and EPA registration number. Proper application should be validated periodically through end-product testing for chemical residue levels. Microbial testing also may be necessary for treated process water.

Plant Pesticide Control

When rodenticides, avicides, and insecticides are applied at processing facilities, strict controls are needed to prevent product contamination. These materials should be stored in a locked, well-ventilated storage area with the proper signage. The materials should be labeled as approved for food plants, and sample labels and SDSs should be on file. Rodenticides should be applied only to exterior areas of the facility in locked, tamper-proof bait stations. All containers of pesticides should be properly labeled. Special keyed bait stations, bait-securing devices for bait stations, and employee safety programs are a few of the best practices and new technologies in the area of plant pesticide control.

Maintenance Chemical Control

All lubricants used in process zones should be certified as food grade. Application of lubricants should be carefully controlled to prevent product contamination. Lubricants should be applied only during line downtime, and there should be a written standard operating procedure for proper application techniques. Bearings, hydraulic drives and lines, and conveyor drives should be located outboard to process lines, and catch pans should be installed under bearings located directly above process areas. Lubricants should be labeled and stored in the same manner as pesticides and sanitation chemicals. If non-food-grade lubricants are used in the process zone, these chemicals should be stored in a secured location that is separate from the area used to store food-grade lubricants.

A spill control procedure should be developed for lubricants and other maintenance-related chemicals (such as solvents and anhydrous ammonia). Forklift battery charging areas should not be located near process zones and should be designed with spill containment barriers. Best practices and new technologies in the area of maintenance chemical control include spill "pigs" (containment devices), greaseless bearings, and employee safety programs.

Allergen Control

Allergen control concerns chemical hazards attributed to raw materials and ingredients. An allergen control program should be established when manufactured food products have allergenic ingredients. Such a program evaluates and controls the risks from allergen cross-contamination and mislabeling by addressing product design, manufacturing, and packaging processes. Key components of an allergen control program include screening and control strategies.

The screening process involves assessing all ingredients, raw materials, process aids, and packaging to determine if any of these items will induce allergic or chemical sensitivity reactions. Allergenic materials include, but are not limited to, peanuts, tree nuts, milk, eggs, soy, wheat, fish, shellfish, celery, sesame, yellow #5, sulfites, monosodium glutamate, and lactose.

Consideration must be given to the specific regulatory requirements in the geographic origin of the raw material, in the country of manufacture, and in the country of sale. This is necessary to ensure that all the relevant allergens are managed and none are inadvertently omitted due to differences in the regulation. As an example, the list of substances that are considered to be allergenic is different in Canada, Japan, Europe, and the United States.

A documented inventory or registry of these items will demonstrate the need for or exemption from implementing an allergen control program. If allergens or allergen-like materials are processed in the facility, the inventory should identify on which process lines these materials are run and if the lines are stand-alone or are shared by non-allergen-containing products. A finished product inventory that lists all finished food products containing allergens and identifies which allergen(s) each product contains also should be developed. Such an inventory is an important tool in planning the sequence of production runs to avoid allergen cross-contamination.

Numerous control strategies can be used to manage allergens, depending on the type of allergenic ingredients/raw materials and type of finished food product. The starting point is a hazard analysis, which should be used to identify all allergen food hazards in raw materials, ingredients, the processing system, and packaging. The hazard analysis will assess the likelihood and severity of each potential allergen hazard. It will also distinguish those significant hazards that should be controlled through a HACCP system from hazards that can be controlled through PRPs. In some cases, a typical PRP, such as cleaning and sanitation, may be elevated to a CCP for allergen control. This occurs when the hazard analysis has determined that failure to clean equipment or surfaces properly could lead to a life- or health-threatening condition.

Where non-allergen-containing products are run on the same process lines as allergen-containing products, the hazard analysis should evaluate the effectiveness of cleaning after allergen-containing product runs and also should include testing of the non-allergen-containing products for cross-contamination from allergens.

Receiving controls should be in place to address allergens. These include certification of incoming materials as allergen-free, labeling incoming pallets of allergenic ingredients, segregated storage of allergen-containing materials, and review of incoming packaging materials to ensure ingredient listings include allergenic ingredients where required.

There are many control steps for allergens during processing and packaging. Where possible, allergen-containing products sharing the same process line as non-allergen-containing products should be scheduled to be run last, followed by a full cleanup. Dedicated utensils, containers, tools, uniforms, and scaling equipment should be used for allergen-containing product runs. It is critical that all allergenic raw materials and ingredients be listed properly in the ingredient declaration on packaging material. This can be monitored through manual operator checks and by automated optical scanning of universal product codes (UPCs) on packaging material that automatically rejects product or shuts down the manufacturing line if allergen-containing product is detected.

Cleaning is one of the most important controls for allergens. Procedures for allergen cleanups should be thorough and documented. Post-cleaning visual inspection is critical to ensure that all allergenic residue is removed from process lines; inspection may be augmented by ATP bioluminescence swabbing or swabbing for allergens.

Lines must be properly engineered to control allergen cross-contamination. Allergenic ingredients should be added at the furthest point in the process flow toward packaging. Lockout of three-way valves and allergen applicators should be used on shared product lines. A dust removal system should be installed to filter finely divided allergenic materials (such as peanut flour). Traffic flows from allergen scaling areas into process areas should be carefully evaluated.

Training is key for all of the above controls. Employees need to be aware of allergen control procedures, and training should be conducted for new hires and on an annual refresher basis. As always, results of training should be documented.

Where the same process line is used to run allergen-containing and non-allergen-containing product, the allergen control program must be validated by end-product testing. "First-off" product from non-allergen-containing food product process lines should be tested for the presence of cross-contaminating allergens on a routine basis.

Best practices and new technologies in the area of allergen control include the labeling of incoming allergenic ingredients, reviewing allergens at receipt, allergen warnings on product labels, UPC scanning during packaging to ensure the correct packaging material is used for allergen-containing products, chemical labels and SDSs, dust removal systems for areas handling powdered allergens, lockout for allergen application equipment, allergen mapping, and rapid enzyme-linked immunosorbent assay (ELISA) test kits for egg, dairy, and peanut allergens.

Cleaning and Sanitation

Cleaning and sanitation are important prerequisites that deal with the housekeeping, cleaning, and sanitizing procedures used to control possible contamination in the manufacturing facility. Cleaning generally refers to the removal of soils, debris, and chemicals from food processing equipment and environmental surfaces. Sanitizing, on the other hand, is the application of microbicides to cleaned surfaces for the purpose of killing microorganisms. The sanitation process follows the cleaning process.

Key components of a cleaning and sanitation program include a master cleaning schedule, a daily housekeeping schedule, written cleaning procedures, and housekeeping practices. A master cleaning schedule is a tool used to ensure that other-than-daily cleaning tasks are completed on a timely basis. It identifies all key cleaning tasks for equipment, outside grounds, building areas, and food utensils; the required frequency of cleaning; and the responsible position or person; and it provides a means to track completion dates and employee sign-off. A daily housekeeping schedule is used to inventory tasks that must be completed routinely in plant areas to ensure they are clean, safe, and orderly. This also includes "on-the-run" cleaning. The overall goals of using cleaning schedules are to: (1) provide a method to manage a large number of important tasks that cannot be practically assigned to memory; and (2) schedule cleaning activities with a frequency that will disrupt the life cycles of insects and microorganisms.

Written cleaning procedures are used to document the sanitation program. These are work instructions that detail how to clean plant equipment, what types of chemicals are necessary, procedures for using application equipment and other cleaning devices (high-pressure guns, low-pressure hoses, clean-in-place [CIP] tanks, clean-out-of-place [COP] tanks, foaming vessels, line cleaning air-actuated bullets ["pigs"], and so on), and safety procedures that need to be practiced when handling chemicals and equipment (such as personal protective equipment and lockout/tag-out). A cleaning procedure should be developed for each key piece of food processing equipment, as well as for the food processing environment (floors, walls, ceiling). Procedures should be written in simple language, preferably by the personnel who will be performing the actual cleaning and sanitation.

Best practices and new technologies in the area of cleaning and sanitation include the use of color-coded brushes, utensils, and other cleaning equipment to prevent cross-contamination between raw and finished product work areas; peracetic acids for equipment sanitation; use of digital cameras to document cleaning procedures; training programs for sanitation crews; automated titration and tracking of chemicals during CIP and COP cleaning; automated belt washers; a high-pressure cleaning attachment ("mouse") for pipe washing; rapid results ATP bioluminescence swabbing equipment for monitoring cleaning effectiveness; and chemical spill pallets.

MICROBIOLOGICAL CONTROL

Microbiological control involves a program to monitor, assess, and control the risk of microbial contamination. Typically, cleaning and sanitation address control of microbiological hazards, but this additional prerequisite is needed for facilities where control of such hazards as listeria and salmonella is critical.

Key methods of microbial verification include equipment swabbing, line profiling, environmental monitoring, and product testing for microbiological contamination. Swabbing is used to evaluate the effectiveness of cleaning and sanitizing on process equipment and to problem-solve product contamination issues. Line profiling involves product sampling at different locations along the process flow to determine sources of microbial loading in food products. Samples of raw materials, intermediate product from various process steps, and finished product are tested for total plate count, yeast and mold, coliforms, and specific pathogens to build a microbial profile of the process system. Environmental monitoring is the central component of a *Listeria monocytogenes* control strategy and involves sampling surfaces in areas where cooked, refrigerated, or perishable products are processed and packaged. These surfaces include floors, walls, ceilings, drains, trash cans, trolleys, conveyor frameworks, storage hoppers, sinks, HVAC equipment, and other potential environmental sources of pathogens. Where food products are cooked or there is a pathogen kill step, it is especially critical to monitor areas between the kill step and packaging. Salmonella monitoring may be conducted in areas where products with a history associated with salmonella are used, such as dried dairy products, eggs, and meat.

Other important components of microbial control include self-inspections and audits, hand washing, hand dipping, foot baths, control of plant traffic, sanitary design of process lines, and clean gloves and uniforms. Vigorous hand washing must be performed for a minimum of 20 seconds with hot, soapy water to remove most microorganisms. Many of these items are part of other PRPs. Finally, microbial testing and timely corrective action in response to issues identified during audits ensure pathogens are controlled or eliminated.

Best practices and new technologies in microbiological control include rapid swabbing test kits; sponge sampling kits for listeria; floor foams; color-coded cleaning equipment and maintenance tools; Microban-impregnated paints, conveyor belts, and toilet and door handles; and drain sanitizers.

Sanitary Design and Engineering

Production facilities and process lines should be designed to prevent contamination of food products. Proper design impacts many areas: the building, equipment, electrical systems, construction, maintenance, cleaning, solid waste handling systems, pest control systems, and foreign material control.[9] Sanitary design criteria depend on the specific industry segment and inherent food hazards (such as dairy processing versus flour milling), although certain engineering requirements pertain to all food plants. Food defense should be considered when planning or auditing a facility. An assessment of the potential risks of intentional adulteration should be completed, and mitigation strategies should be implemented when necessary.

Each facility or food company should develop a set of sanitary design standards to serve as a guide for constructing new process lines and production facilities. These industry-specific guidelines also are an important training and compliance tool for company maintenance personnel and outside contractors. For example, guidelines provide directives for the use of sanitary welds during equipment installation and line construction. Ground and polished sanitary welds ensure equipment and surrounding surfaces can be fully cleaned and sanitized, thus augmenting microbial control.

These universal requirements relate to basic building design and location. Plants should be located away from feedlots and landfills. Drainage from feedlots or landfills should be directed away from the processing site. Adequate drainage and dust control must be incorporated into outside grounds through drains and paved lots. Landscaping should not attract birds, insects, rodents, or other pests;

low shrubs and grass are preferred over trees, but shrubs should not be planted next to buildings. Lighting should not be mounted on the building, but rather on poles or other fixtures that illuminate the building from a distance and draw insects away from the facility. Design features incorporated into the facility to prevent rodent entry include outside drainpipe screens, metal flashing installed below loading docks, door seals, and a graveled perimeter.

The interior of the plant should be designed to facilitate cleaning. Equipment should not rest directly on the floor. Services such as pipes and electrical conduits should be mounted away from walls. Where applicable, an equipment loft should be incorporated into the plant design so utilities such as steam, air, power, and water can be supplied to process equipment from directly overhead.

The need for environmental control and monitoring should be assessed for the various programs within the system including control of temperature, humidity, dust, pathogens, water, air and ice safety, and facility design elements.

Air handling systems must be designed to prevent contamination. Generally, positive pressure is desired in process areas, and air should flow from finished product areas to raw material areas, not vice versa. Ductwork may need to be insulated to prevent condensation buildup over process zones.

Restrooms and hand-washing areas are a regulatory requirement. Hot and cold water must be supplied to hand-wash stations, and valves should be foot or electronically activated to prevent recontamination of hands. Restrooms must not open directly into production areas.

Raw material preparation areas should be segregated from intermediate and finished product processing areas. Design should take employee and forklift traffic flows into account to minimize the potential for cross-contamination.

Best practices and new technologies in sanitary design and engineering include 3-A Standards, National Sanitary Foundation (NSF) design standards, Baking Industry Sanitation Standards Committee (BISSC) design standards,[10] EU design standards,[11, 12] Foundation for Meat and Poultry Sanitary Equipment Design Principles, antimicrobial additives for paints and plastic materials, utility lofts, and a sign-off process for new equipment and process lines.

Preventive Maintenance

Preventive maintenance involves the use of a predetermined schedule to service the physical building, equipment, and processing utensils with the goal of preventing food product contamination. This prerequisite ensures that structural beams, supports, walls, ceiling, and floors are maintained on a regular basis to eliminate contamination from paint chips, insulation, metal, plastic, or wood. Overhead light fixtures must be properly maintained to ensure they are adequately shielded against glass breakage. Equipment such as conveyor belts, bearings, drive motors, chain guards, augers, and pumps must be serviced on a regular basis to prevent contamination from leaking lubricants, conveyor clips, belt threads, rubber gaskets, and metal shavings from metal-on-metal wear.

The core of a good preventive maintenance program is a schedule and a work order system. The schedule is a management tool that ensures equipment and structures are routinely serviced before they become a source of contamination.

The work order is used to track scheduling and completion of both preventive maintenance and non-scheduled repairs. Work orders should include a priority system that gives an urgent status to food safety-related repairs. A good preventive maintenance program also should address the removal of project debris after completion of work by maintenance personnel or contractors.

Best practices and new technologies in the area of preventive maintenance include predictive maintenance, computerized preventive maintenance systems, contractor food safety requirements, and total productive maintenance systems.

Pest Control

Pest control involves a program to limit pest activity through documented programs and practices. Key target pests include birds, rodents, and insects. A written pest control program should be on file, describing practices used to control birds, rodents, and insects. The program should include the following documents: overview of the program, current applicator license and liability insurance for the pest control operator (PCO) or in-house applicator, written procedures for chemical application, sample labels and SDSs for all pesticides, a schematic showing the location of all pest control devices, service reports for the PCO, and a pesticide application log. If restricted-use pesticides are used, the applicator must be properly licensed; the application of general-use pesticides typically requires a trained applicator (requirements may vary, depending on local and state laws).

Controls to limit rodent activity may include exterior bait stations, interior mechanical traps, glue boards, hardware cloth, door seals, and graveled or paved zones around the plant. Rodent activity should be limited by proper sanitation to remove food debris and to eliminate harborage areas.

Insect controls may include insect light traps, door and window screens, pheromone traps, glue boards, fumigation, fogging, and spot spraying with insecticides. Sanitation is also an important preventive measure in controlling insects, and efforts must be made to remove sources of food, water, and harborage.

Bird control typically is addressed through a combination of strategies. The first approach should be removal of all sources of food and sites for roosting and nesting. Tools used to facilitate bird control include gang spikes, plastic owls, predator balloons, hardware cloth and bird nets for exclusion, pellet guns, screened windows and doors, and avicides.

Best practices and new technologies in pest control include risk-based practices defined in the 2016 Pest Management Standards For Food Processing & Handling Facilities, National Pest Management Association,[13] pest findings trend analysis, use of personal digital assistants to scan pest control devices (electronic logging), monitoring with nontoxic bait blocks, pheromone monitoring, CO_2 fumigation, tin cat mechanical traps, 18–24 inch (0.5–0.67 m) inspection aisles, 30–36 inch (0.75–1 m) gravel perimeter, exterior bait stations every 50–100 feet (15–30 m), interior mechanical traps every 20–25 feet (6–8 m) and flanking doors, non-electrocuting insect light traps (glue boards), 30-foot (9-m) buffer zones between electric grid insect light traps and exposed process zones, annual changeout of insect light trap tubes at peak insect season, solar fly traps, and parasitic wasps for outdoor fly control.[13]

Receiving, Storage and Shipping Controls

Food products must be handled in a safe and sanitary manner during the receipt and storage of raw materials and the storage and distribution of finished food products. Receiving, storage, and shipping controls utilize numerous practices to prevent product contamination at the beginning and end of the food manufacturing process.

Receiving controls include inspection of all incoming carriers for sanitary condition, inspection of all incoming ingredients for potential contamination, proper documentation of all incoming raw materials and ingredients, temperature evaluation of all incoming perishable raw materials, and proper documentation of incoming product safety, such as certificates of analysis (COAs). Results of incoming material inspections should be properly documented rather than reported by exception. Tracking should include date, supplier, lot number, temperature (if applicable), condition of carrier, evidence of tamper seal on carrier, carrier identification, and condition of product and pallets. COAs should arrive prior to or with incoming shipments and must be examined to ensure compliance with company written specifications.

Company written specifications typically are developed for all raw materials, packaging, ingredients, process aids, and finished food products. Included in a specification are a product description, transportation and storage requirements, and required analytical test results for quality attributes and known biological, physical, and chemical hazards. In addition, some operations pull samples of incoming materials for acceptance testing of microbiological, visual, physical, and other product attributes.

Once received, all raw materials should be dated directly on their containers rather than on shrink-wrap to ensure first-in first-out (FIFO) utilization. It also has become common to specially identify pallets containing allergenic ingredients. Raw materials and ingredients should be stored in areas segregated from processing and packaging areas, and allergen-containing ingredients often are further segregated. Holes and tears in ingredient containers created during the unloading or storage process should be inspected for contamination, and either the product should be discarded or the damage should be repaired by taping and labeling tears. Perishable and frozen ingredients must be placed in storage at the appropriate temperature, and temperatures should be monitored and recorded on at least a daily basis.

Storage areas must be maintained in sanitary condition. This requires appropriate cleaning and pest control measures. Palletized ingredients and raw materials should be stored at least 18 inches (0.5 m) off the floor, preferably in racks, and should be at least 18 inches (0.5 m) away from walls and ceilings to allow for maintenance aisles. Where it is necessary to store goods on the floor, slip sheets should be used as a sanitary barrier.

Programs for finished goods are parallel to those for incoming ingredients. Food products should be stored in a safe manner in clean areas with appropriate pest controls. A segregated sanitary storage area typically is established for damaged goods; these areas need daily scrutiny to manage spillage and potential food sources for pests. The temperature of refrigerated and frozen goods must be routinely monitored and logged to prevent quality and food safety issues. As in receiving areas, storage of finished food products must be separate from storage of chemicals, including food-grade chemicals used as ingredients or process aids. Outgoing carriers must also be inspected to ensure they are free from odors, toxic chemicals, debris, foreign materials, rodents, and other hazards. Company-owned distribution vehicles should be cleaned regularly to ensure sanitary condition.

Best practices and new technologies in receiving, storage, and shipping controls include the examination of shipping records for the prior three cargoes transported by outgoing carriers, color-coded FIFO pallet tags, and allergen labeling.

Supplier Controls

Supplier control refers to a program of company criteria for the evaluation and approval of suppliers, raw materials, ingredients, and services to minimize food product contamination. Without effective supplier control, even the best prerequisite and HACCP systems cannot fully ensure product safety.

The starting point is clearly defined expectations for suppliers, usually in the form of specifications and a supplier approval checklist. Ingredient specifications should outline requirements for control of food hazards in supplied materials. Compliance to these specifications is usually demonstrated in COAs accompanying each shipment (see previous section).

It has become standard practice to establish a supplier approval program to further protect the customer from contaminated raw materials. Typical expectations for suppliers include an implemented HACCP program, participation in a third-party sanitation audit program with a favorable audit rating, a product liability insurance policy, a documented trace and recall program, a continuing pure food guarantee of product safety, and an on-site qualifying food safety and quality audit conducted by a customer representative.

The formation of partnerships and alliances, the identification of select supplier programs, Internet accessibility of supplier specifications, and supplier performance review meetings are a few best practices in the area of supplier control.

Water Supply

Although water safety is addressed in the GMPs, the existence of numerous issues with potability and contamination of drinking water with pathogens such as *E. coli O157:H7* is justification for putting more emphasis on this important prerequisite. A water safety program mandates a process to manage the safety and quality of water used as a food ingredient, water used in processing, water used in cleaning operations, ice and steam used for food contact, and drinking water consumed by employees.

Only potable water can be used by food processing operations for the purposes listed previously. An annual Consumer Confidence Report for water quality from your local source that verifies compliance with the EPA National Primary Drinking Water Regulations should be maintained on file for water used in the facility. For water supplied by a municipality, documentation of compliance typically is supplied by a water department or company upon request. When well, lake, or river water is used by a food company, the water must be treated and tested by the company on a routine basis to certify compliance to EPA potability requirements. In addition to meeting EPA requirements, non-city water should be microbiologically tested on a weekly basis to document the effectiveness of water treatment systems and freedom from coliforms and pathogenic bacteria (and in some cases, protozoa). Where water is a predominant food product ingredient (such as in beverages), further written specifications should be developed to outline other important quality-related criteria, such as hardness, off odor or flavor, chlorine levels, and particulate levels.

Water treatment must be carefully monitored and logged by the processing facility. Daily testing of chlorine gas, calcium or sodium hypochlorite, or ozone levels is necessary to monitor correct chemical application levels and desired microbial kill. Nonchemical treatments such as ultraviolet or heat pasteurization systems also need to be carefully monitored and logged to ensure correct operation. Where food-grade steam is utilized, careful management of boiler treatment chemicals is necessary to prevent product contamination. Personnel should ensure that chemicals are labeled for food-grade use, that they are added at the appropriate concentrations, and that chemical labels and SDSs are maintained on file.

Food plants must be properly engineered to prevent backflow or siphonage of wastewater, wash water, septic lines, or other non-potable water into potable water lines. Backflow devices should be installed on potable water lines used for drinking water, cleaning water, ice production, and processing water supply. Plumbing must be done to eliminate dead legs (or back legs), removing areas where microorganisms can grow. Installations must comply with federal, state, and/or local regulations. Backflow devices should be inspected periodically by maintenance personnel or third-party agencies.

If a site uses gray water, the gray water distribution system must be a separate water supply system and not connected to the potable water system.

Best practices and new technologies in water safety include filtration of all ingredient water with a 10-micron filter, backflow blueprints, and annual backflow certification.

Air and Gas Supply

Air safety describes a program to manage the safety and quality of air and gases used in a facility HVAC system, as a food ingredient, in processing, in cleaning operations, and in packaging operations. Failures related to this program can result in such hazards as airborne microorganisms, airborne peanut protein, hydrocarbon contamination from air used in packaging machines, and chemical contamination of carbon dioxide used in carbonated beverages.

Air either used in the HVAC system or supplied to equipment such as dryers and classifiers should be filtered to remove dust, insects, and other small contaminants. For microbiologically sensitive food products, high-efficiency particulate absorbing (HEPA) filtration may be necessary. Filters should be cleaned or replaced on a regular basis, managed through the preventive maintenance or master cleaning schedule. Evaporative coolers also need regular cleaning and maintenance to prevent microbial growth in the water spray system and on evaporation media units.

Gases and air used to process and package food products must be food grade to prevent contamination. Toxic lubricants, dirt, water, and other materials must be removed from compressed air that contacts food products. Compressed air should be filtered to remove particles of at least 50 microns in size. Traps and filters require maintenance on a regular basis and results should be logged.

Gases such as nitrogen and carbon dioxide that are used as ingredients or process aids must be food grade. Specifications from suppliers should stipulate this status. Based on risk, a COA may be requested with each incoming shipment.

Air in specific processing rooms needs to be monitored to minimize the risk of cross-contamination. In many cases, specified air pressure differentials need to be maintained between processing rooms. Airflow in the plant should move from "clean" areas to "dirty" or contaminated areas.

Best practices and new technologies in air safety include microbiological air sampling, the use of charcoal and HEPA filters, UV lights in the cooler ducts, and the use of food-grade gases.

Food Safety Training

Many PRPs have a training component. This includes training in GMPs, cleaning and sanitation, personal hygiene, allergen control, food defense, and preventive maintenance. Due to the extensive amount of training required to effectively manage PRPs and HACCP, it is key that a system be established to provide for the scheduling, presentation, and tracking of employee training.

Training formats can be group presentations or individualized learning from interactive programs. An important part of training programs is identifying appropriate educational materials from the wide variety of videos, interactive CD-ROMs, manuals, and other commercially available material.

Upon completion of all training sessions, employee learning should be evaluated and results documented. Documentation of training must include the date, subject matter, name of student, title of subject matter, and results of written or oral subject-matter testing, along with a roster of all employees attending the training session. Training records should be retained in the employees' personnel files or other appropriate locations.

A calendar or master schedule is a useful tool for ensuring timely completion of required training. Typically, food safety training is required for all new employees, with annual refresher courses given in such areas as cleaning practices and GMPs.

Best practices and new technologies in food safety training include computerized tracking programs for completed training, interactive online food safety training programs, and the establishment of learning centers for individualized training.

Equipment Calibration

Equipment calibration involves the standardization and calibration of analytical and processing equipment used to control food safety hazards. Thermometers used to track microbial kill temperature in food products must be calibrated on a regular basis to ensure reliability. Scales used to weigh food additives that have regulatory tolerances should be standardized with certified weights on a frequent schedule. Examples of other equipment that may require calibration include pH meters, moisture analyzers, vacuum gages, and micrometers.

Where possible, a certificate should be obtained through a national or international standards organization for calibration devices such as standardized thermometers and metal detector test balls. Calibration should be performed on a scheduled basis and results documented. Standard operating procedures typically are developed to outline the approved process for performing equipment calibration. The use of calibration management software and calibration labels is a highly recommended practice.

Customer Complaints—Food Safety

This PRP involves the review of marketplace risk associated with customer complaints, and investigation and corrective measures needed to prevent recurrence. The program focuses only on product safety-related complaints and must have a mechanism to identify complaints, trends, and frequencies that may make product withdrawal or recall necessary.

A written program should be developed to describe procedures for handling food safety complaints, including forms for tracking the complaint from receipt through resolution. Most companies share complaint information with employees to aid in rapid troubleshooting and corrective action to eliminate the root cause of product contamination.

In many cases, investigation of complaints may require analytical capabilities to identify the product contaminant properly. Examples of testing include analysis for microbial pathogens, glass flame and chemical tests, foreign material analysis, insect analysis, hair identification, and chemical testing.

The following are excellent methods ot tracking complaints: computerized complaint tracking programs, complaint trend analysis, complaint log books, complaint measurement, and normalization of data by units produced/sold/ distributed.

Audits and Inspection Programs

PRPs require ongoing management and evaluation. Audits and inspections are used to evaluate the effectiveness of these programs and must be utilized to ensure constant improvement of the control of food safety hazards. Audits identify program defects, verify systems are in place, and are a starting place for corrective actions. Audits and inspections include monthly food safety self-inspections, pest control inspections, post-cleaning inspections, GMP inspections, third-party food safety and sanitation audits, and supplier approval audits.

The starting point for an inspection or audit should be a standard, guideline, or standard operating procedure that defines expectations for the area, system, or equipment under review. For example, company GMP and personal hygiene rules for employees outline requirements that will be evaluated during the plant floor GMP inspection.

Results of audits and inspections should always be documented. Upon completion of the review, results should be recapped with the appropriate personnel or work group. Opportunities and defects should be prioritized for corrective actions.

In some cases, a score or measure may be developed from an audit. This measure of program effectiveness should be communicated to all personnel, from line workers to executive management, to facilitate corrective actions, to acquire needed resources, and to identify progress made in food safety and sanitation programs.

Best practices and new technologies in auditing and inspection include the use of personal digital assistants and handheld computers for plant inspections and the use of food safety teams for management of self-inspections and corrective actions.

NOTES

1. *Code of Federal Regulations,* Procedures for the Safe and Sanitary Processing and Importing of Fish and Fishery Products (Seafood HACCP Regulation), title 21, part 123 (1995).
2. *Code of Federal Regulations,* Hazard Analysis and Critical Control Point (HACCP) Systems, title 9, parts 416 and 417.
3. U.S. Food and Drug Administration, *Guidance for Industry: Guide to Minimize Microbial Food Safety Hazards of Fresh-Cut Fruits and Vegetables* (Washington, DC: Center for Food Safety and Applied Nutrition, 2008).
4. AIB International, *AIB Consolidated Standards for Food Safety* (Manhattan, KS: AIB International, 1995).
5. AIB International, "HACCP Overview," in *HACCP Workshop* (Manhattan, KS: AIB International, 2000).
6. William H. Sperber, et al., "The Role of Prerequisite Programs in Managing a HACCP System," in *Dairy, Food, and Environmental Sanitation* 18, no. 7 (1998): 418–23.
7. National Advisory Committee on Microbiological Criteria for Foods (NACMCF), *Hazard Analysis and Critical Control Point Principles and Application Guidelines,* Appendix A (Washington, DC: U.S. Food and Drug Administration, August 14, 1997).
8. U.S. Food and Drug Administration, "Foods Adulteration Involving Hard or Sharp Foreign Objects," in *FDA/ORA Compliance Guide,* Chapter 5, Subchapter 555, Section 555.425 (1999).
9. Thomas J. Imholte and Tammy Imholte-Tauscher, *Engineering for Food Safety,* 2nd ed. (Woodinville, WA: Technical Institute for Food Safety, 1999).
10. Baking Industry Sanitation Standards Committee (BISSC), *1998 Sanitation Standards for the Design and Construction of Bakery Equipment* (Chicago: BISSC, 1998).
11. European Committee for Standardization, *Food Processing Machinery—Basic Concepts—Part 2: Hygiene Requirements* (British Standard EN 1672-2: 1997, Subcommittee MCE/3/5, Food Industry Machines. English version, 1997).
12. Foundation for Meat & Poultry Sanitary Equipment Design Principles, *2014 Sanitary Equipment Design Taskforce.*
13. National Pest Management Association, *2016 Pest Management Standards For Food Processing & Handling Facilities.*

Chapter 3

Other Food Systems Support Programs

To round out foundational control programs, some additional systems are necessary. There is some debate among experts as to whether a program is a prerequisite program, a control program, or something new altogether. Care should be taken not to be too concerned with where to classify these support programs, but to make certain they have been considered and addressed. To be clear, you must have a clean operation and have addressed the programs discussed below and in the previous chapter if you have any hope of your HACCP system, and ultimately your food safety management system, to be in control.

PRODUCT TRACEABILITY AND RECALL

Trace and recall is a program used to track and control the movement of food products from receipt of ingredients and raw materials to end-point distribution of finished goods. This level of control is established to enable a food company to retrieve product from the distribution system and marketplace in the event a product is defective or becomes contaminated during manufacturing or retailing. The elements of a good recall program include a written product withdrawal and recall policy that includes a defined recall process; a recall action team; proper lot coding of all retail and food service packaged units; product complaint handling procedures; a system of notification for company personnel, customers, and regulatory agencies; a means to recover the food products; and a means to dispose of the food products properly to prevent them from reentering the food chain, and may be included in the firm's crisis management plan. It is essential that all contact information be up to date.

Facilities should maintain accurate records of the lot or batch numbers assigned to food products. Lot or batch numbers should be incorporated into distribution documents such as shipping manifests or bills of lading to facilitate product tracking, and copies of these records should be held based on regulatory and consumer requirements for at least the shelf life of the product. Food companies should periodically test the effectiveness of their trace and recall program through mock recall exercises. A mock recall is intended to verify the recall plan's effectiveness while demonstrating the firm's ability to identify both the affected product and the status of that product within the chain. Label control exercises can also be used to verify the effectiveness of the firm's label controls. The results of these exercises should be summarized, documented, and maintained on file.

Trace and recall procedures include tracking lot numbers of raw materials to specific product lot numbers, ensuring rework is tracked into product lots, production records, mock recall exercises, computerized warehousing and recall tracking, company spokespersons, crisis management plans, preplanned press releases, disposal certification, monitoring of coding equipment, and printed lot codes. It is essential that the organization be able to forward- and backward-trace raw materials and packaging to finished product lots. At a minimum, the company should be able to trace raw materials and packaging from the immediate supplier to finished product to the next customer. This is sometimes called "one-down, one-up."

Rework should be clearly identified and/or labeled to allow traceability. Traceability records for rework should be maintained. The rework classification or the reason for rework designation should be recorded (product name, production date, shift, line of origin, shelf life, and so on).

CRISIS MANAGEMENT

Firms should assess the risk of crisis that may occur in their system with consideration to both environment and operations. Consideration of power outages, natural disasters, mass scale contamination, public health outbreaks, and recall capabilities are examples. Emergency preparedness plans may also be included in an effective crisis management plan. An effective crisis management plan ensures that all personnel within the food safety system have been provided training to ensure they know what to do when an incident or crisis event occurs. The outcomes of an effective crisis management plan mitigate the consequences of crisis events.

FOOD DEFENSE AND FACILITY DESIGN

Food defense, the consideration of intentional adulteration and contamination, has been a growing concern over the last two decades. This requires thought by the firm as to how to protect the food supply from those who desire to cause pain and disruption. Governments are concerned, of course, from the terrorism point of view, but a firm must also consider possible threats from its competitors designed to disrupt the business in order to gain an economic upper hand, as well as harm desired by disgruntled employees. Guidance is available, such as the FDA's FSMA Final Rule Mitigation Strategies to Protect Food Against Intentional Adulteration. Food defense plans usually contain elements similar to a HACCP plan such as a vulnerability assessment, mitigation strategies, monitoring, corrective action, verification, training, and record keeping.

ENVIRONMENTAL CONTROL AND MONITORING

Many firms also combine some of the prerequisite programs described into a specific program called environmental controls. Often this is in response to a specific need, such as a listeria control in a food plant producing ready-to-eat foods without a heat treatment. Such controls combine sanitation treatments and procedures, monitoring of areas such as air and gas supply, and sampling and testing to infer control overall of a specific concern, usually microbiological. These systems also require proper training and documentation.

Environmental monitoring is used to verify the control programs designed to significantly minimize or prevent environmental pathogen contamination of ready-to-eat foods are working effectively. Sanitation may not be the only control necessary to prevent recontamination of exposed ready-to-eat foods, especially when raw and ready-to-eat products are produced in the same facility.

Food Fraud

Food fraud, while not considered a necessary standalone PRP, must be considered when evaluating any food safety system. The risk of intentional or unintentional use of ingredients (for example, substitution, mislabeling, misbranding, dilution, and counterfeiting) that may compromise economic integrity, safety, and/or quality of the final product must be evaluated and prevented. Such considerations must include the supply of raw materials and possible fraud occurring after shipment.

Threats relating to the adulteration or substitution of raw materials constantly change. These threats must be monitored and identified and existing ones managed. Sites are to remain up to date with emerging issues and must adapt systems to protect their products against new and existing threats.

Food fraud plans usually contain elements such as a vulnerability assessment, mitigation strategies, monitoring, corrective action, verification, training, and record keeping.

Chapter 4

Tasks for HACCP Plan Development

ASSESSING THE NEED FOR A HACCP PLAN

The information contained in a HACCP plan will vary since unique cultural issues and many different processes exist within individual companies. Normally a HACCP plan is product- and process-specific, but some plans use a unit operations or recipe approach. For example, in a retail setup, a HACCP plan could be developed for a specific clam chowder recipe or for heat-processed foods in general.

Assessing the need for the implementation of a HACCP plan is the responsibility of the executive management group. External pressures to implement a HACCP system are exerted on industry by two primary sources: government regulations and customer requirements. In industries in which HACCP is mandated by government (regulatory HACCP), the choice to implement and maintain a viable HACCP system is a foregone conclusion because it is a requirement of doing business for both large and small organizations. Although less prescriptive, customer-motivated HACCP requirements typically are viewed as being market-driven or as offering a strategic advantage in a competitive marketplace. The following is a list of common reasons for implementing a HACCP program:

- The company's internal nonconforming product is responsible for the loss of a significant sum of money
- Competitors making similar products have experienced marketplace failures that have resulted in costly product recalls, loss of customers, and loss of market share
- National and international government agencies and standards-setting groups require all processors, distributors, and retailers to participate in a regulated HACCP or food safety program
- A large customer mandates that its suppliers must implement a verifiable HACCP program to remain a preferred supplier
- Even when a HACCP program is not required, many companies voluntarily choose to implement one because they think it is the right thing to do and believe it may provide a marketing advantage[1]

PRELIMINARY TASKS FOR HACCP PLAN DEVELOPMENT

In the development of a HACCP plan, five preliminary tasks need to be completed before the HACCP principles are applied to a specific product and/or process: (1) assemble the HACCP team; (2) describe the product and its distribution; (3) describe the intended use and consumers of the product; (4) develop a flow diagram that describes the process; and (5) verify the accuracy of the flow diagram.

Assembling the HACCP Team

The executive management group is responsible for providing the necessary budget and resource planning to ensure effective implementation and maintenance of the HACCP system. When communicating the need for HACCP and expressing the desire to make it part of the organization's culture, management should clearly define the goals of the program and determine when the program is expected to be fully operational. Some companies include their product safety goals in policy statements. These statements should be easily understood by all company personnel. An example of a typical policy statement is "To produce safe product worldwide."

An upper-level manager often signs the HACCP plan as a record of official endorsement. The executive management group is responsible for communicating both the direction of the organization and the need to change for regulatory compliance and customer satisfaction.

Regardless of the size of the organization, an individual employee's knowledge of product safety issues in raw materials, process, product use, and distribution requirements will be influenced by diverse circumstances and a unique corporate culture. Regulated HACCP systems specifically require that personnel involved in the planning, implementation, and maintenance of a HACCP system receive documented training. To analyze and develop the resources available within the organization, it is recommended that the HACCP team be multidisciplinary. This helps ensure that primary and shared responsibilities are not overlooked or heavily loaded onto one department, such as the quality department.

Table 4.1 may be helpful in deciding which departments have primary or shared responsibilities for the steps required for HACCP plan implementation. Table 4.1 could be useful in audit planning since it indicates the department and personnel responsible for individual components of the HACCP system.

The task of assembling and maintaining a HACCP team is an auditable activity. A HACCP auditor must allow for differences in approach and company culture when reviewing the structure and participants in an organization's HACCP system. Auditors reviewing a program controlled by a HACCP plan must be open-minded and focused on scientific suitability and effective execution of the product safety plan. Competent auditors exclude personal preferences while conducting audits and report on positive and negative aspects of the program in an objective manner.

Table 4.1 Establishing company accountabilities and audit responsibilities for HACCP.

HACCP requirement	Executive group	R&D	QA	Process and packaging	Purchasing	Sales and marketing	Distribution
HACCP team members	S	S	P	S	S	S	S
Product type and distribution	S	P	S	S		S	S
Intended use and customers	S	P				S	
Develop flow diagram		S	S	P			S
Verify flow diagram		S	S	P			S
1. Conduct hazard analysis		S	P	S			
2. Identify CCPs		S	P	S			S
3. Establish critical limits		P	S	S			
4. Establish monitoring procedures		S	S	P			S
5. Establish corrective actions		S	S	P			
6. Establish verification procedures	S	S	P	S			
7. Establish record-keeping procedures			P	S			

P = primary, S = shared

The company needs to establish which departments and personnel are responsible for the individual HACCP requirements.

The independent HACCP auditors need to establish which departments and personnel are responsible for the individual components of the HACCP system.

A typical audit review of the development of the HACCP team may include the following questions:

1. *Why did the organization start HACCP, and what consultants or company departments were included on the project?* The HACCP team should include balanced representation from all plant departments to ensure personnel with appropriate expertise have participated in the development of the HACCP system. It is not

unusual for the HACCP team to include outside facilitators such as consultants, academics, or the corporate quality group when internal resources are inadequate or unavailable to construct a scientifically valid HACCP plan. If a company decides to use external consultants for the development of the HACCP system, there should be documented evidence that the consultants have the necessary competencies. During the course of the audit, the knowledge level and expertise of HACCP participants should be evaluated by reviewing qualifications, assessing the logic used to construct the HACCP plan, and interviewing employees. The purpose of this review is to evaluate whether the auditee has demonstrated enough cross-functional expertise to adequately analyze the significant biological, chemical, and physical hazards in its product and process. As a group, the ideal HACCP team would include employees with technical and practical knowledge of raw materials, process equipment, packaging, and distribution requirements. The inclusion of manufacturing staff on the HACCP team is encouraged because these employees are the ones who typically monitor critical control points (CCPs). Access to personnel records may be governed by regulatory or company guidelines, so auditors reviewing qualifications should take great care to ensure that the auditee is comfortable with this process.

2. *Have the HACCP team and the team leader or coordinator been suitably identified in the company's documentation system?* The auditor should be able to easily identify the members of the HACCP team, as well as the team leader or coordinator. The HACCP team's responsibilities should be clearly defined in the quality system procedures, work instructions, or forms contained in the HACCP system documentation. The auditor must be able to identify the person(s) responsible for the five preliminary requirements for HACCP plan implementation, as well as those responsible for the application of the seven principles of HACCP: (1) conducting a hazard analysis (Chapter 3); (2) determining CCPs (Chapter 4); (3) establishing critical limits (Chapter 5); (4) establishing monitoring procedures (Chapter 6); (5) establishing procedures for corrective actions and product disposition (Chapter 7); (6) establishing verification procedures (Chapter 8); and (7) identifying records that will be retained as evidence that the HACCP system is effectively implemented (Chapter 9). Well-defined responsibilities for specified requirements, activities, and records to be audited will help the auditors complete their evaluations in an efficient manner. The HACCP team leader has the responsibility of communicating the overall effectiveness of the HACCP system, resolving internal conflict, and communicating resource needs to executive management.

3. *Have the HACCP team and other appropriate personnel received HACCP training?* Personnel responsible for implementing and maintaining the HACCP system should receive initial and ongoing training from an accredited HACCP course provider. If the plant, corporate group, or customer has provided in-house HACCP training, the auditor should review the content of the course to ensure it complies with recognized HACCP guidelines. The auditor should access training records for the HACCP team members, the personnel performing CCP monitoring, and those administering the program to ensure that HACCP training is current.

Potential areas of weakness may be found in companies with high personnel turnover rates or where HACCP systems have been written and implemented with little or no involvement of plant personnel. Where appropriate, an auditor may correlate HACCP system deficiencies to the effectiveness of the training or to the availability of resources provided by executive management.

Auditors should not serve as consultants while performing a third-party audit. However, ISO 19011:2011, *Guidelines for auditing management systems,* allows an auditor to offer nonbinding recommendations termed "opportunities for improvement." These guidelines permit an auditor to inform the auditee where further information, guidance materials, or technical literature can be obtained without compromising impartiality during the audit process. Additionally, the scope of an audit should be communicated prior to the on-site visit and confirmed during the opening meeting to ensure there is agreement as to which products and process lines are to be audited for compliance with the company's HACCP system.

Describing the Product and Its Distribution

The company must have a clear description of the products produced and their distribution requirements, as well as descriptions of any intermediate products or by-products sold as raw materials to other processing plants. This permits the proper identification of hazards and allows the team to reasonably limit the scope of the hazard analysis to events that can occur from manufacturing to the marketplace. The audit team might consider the following questions:

1. *What product(s) does the plant make at this site, and what product line(s) is included in the HACCP system audit?* This simple question will help clarify which areas of the plant and what records are to be assessed during the audit. Once the scope of the audit is confirmed, the audit team can plan the HACCP system audit by assigning areas or criteria to be assessed to individual auditors. It is the auditee's responsibility to provide plant contacts for assisting in the review of HACCP activities and records. The lead auditor typically makes the team auditor assignments, checks on the progress of the audit, confirms nonconformances, and reports on the overall effectiveness of the HACCP system at the closing meeting.

2. *What HACCP system standard is to be applied?* Regulated HACCP systems typically have a defined performance standard and forms issued by the relevant government agency for documenting HACCP plans. The auditors should be familiar with the required forms and have a checklist for the HACCP standard used during the audit. It is common practice for the auditing group to send a copy of the blank audit checklist to the auditee so the audit criteria can be reviewed prior to the actual on-site audit. The auditing group often includes a list naming the auditors who will be performing the audit and their qualifications. This enables the auditee to feel secure that the auditors are qualified to review their product and process.

In unregulated HACCP systems, emphasis and structure may vary because of company and customer influence. Specific customers may have specific performance standards that need to be applied to the HACCP system. It is important to note that the HACCP system standard used should apply only to product safety, not quality issues. The HACCP standard used by the company to formulate its program should be clearly established to explain HACCP system exclusions and potentially conflicting requirements. A reference copy of the HACCP standard should be available for review to define interpretations when differences of opinion occur during the audit. Examples of recognized HACCP systems include, but are not limited to, the National Advisory Committee on Microbiological Criteria for Foods (NACMCF), the WHO/FAO Codex Alimentarius Commission for HACCP, the FDA's Seafood HACCP regulation, and the USDA Food Safety and Inspection Service (USDA/FSIS) HACCP regulation for meat and poultry.

3. *What are the common name, processing methods, and distribution requirements of the product(s)?* In developing the HACCP plan, it is necessary to document the raw materials, ingredients, processing aids, and product contact surfaces. The HACCP plan should list biological, chemical, and physical characteristics, the composition of formulated ingredients including additives and processing aids, origin, method of production, packaging and delivery methods, storage and shelf life, preparation and handling before use or processing, and acceptance criteria or standards related to food safety. In addition, the plan should document the characteristics of the end product by listing the following: product name or identification; composition; biological, chemical, or physical characteristics relevant to food safety; intended shelf life and storage conditions; packaging; labeling related to food safety including instructions for handling, storage, and use; and methods of distribution.

The common name for the product produced typically will be stated in the introduction section of the HACCP system manual, product specification sheet, quality plan, or product form. The auditor should obtain a list of all products produced on site at the plant. This information can be obtained from multiple sources at the plant, such as the sales, quality, or production departments. The information from each source should be compared to evaluate whether the list of products produced is up to date and to verify that new and existing product or process modifications have been effectively communicated to the appropriate personnel. For example, the marketing, product development, and production departments usually are responsible for new products, new processes, process changes, and raw material changes. Has the company effectively reviewed HACCP requirements with regard to new raw materials and the plant's capability to make the product safely on the existing or new equipment? How are changes communicated to the affected departments, and what group reviews their impact on existing HACCP systems? Although changes in raw materials, the process, or product can be viewed as reassessment activities, the organization should explain how these activities are achieved relative to their HACCP system requirements.

The list of products produced can be used to sample the processing methods and testing methods contained in the HACCP plan or the quality plan. The distribution requirements are contained in the quality plan and should include instructions for safe handling of the product to ensure product integrity throughout the distribution chain. Examples of safe handling instructions include labeling the product for shelf life, temperature, and humidity requirements. The testing data and technical information used to determine product handling, storage, and distribution requirements should be available for review to clarify the logic used when establishing requirements for product safety.

Describing the Intended Use and the End User

The normal intended use of the product and likely end user of the product must be clearly defined in the HACCP system documentation. In addition, the HACCP plan should identify any potential abuse or potential mishandling of the product. Even though many companies state that their product is to be used by the general public, certain population groups may have unique risk factors that preclude safe use of the product. Some potential users of the product may have special needs and considerations due to their age or the condition of their health. Typically, pregnant women, infants, young children, the elderly, those with allergen sensitivities, and the immunocompromised present the largest concern because people in these groups may not be able to withstand the stress of treatment with or consumption of the product, resulting in severe health consequences.

Another avenue for reviewing appropriate, or in some cases inappropriate, customer use would be during the review of the prerequisite programs (PRPs) for HACCP (see the "Establishing the Prerequisite Program" section later in this chapter for a general overview of PRPs and Chapter 2 for an in-depth discussion of prerequisite areas pertaining to food safety). Serious safety issues associated with the use of a product should be recorded in the customer complaint program. The auditor should sample customer complaints related to product safety and look for recorded instances where consumption or usage has led to significant illness or injury. If controllable hazards have been identified as a result of consumer complaints, the company needs to institute corrective action and verify the effectiveness of the actions taken.

Developing a Process Flow Diagram

Next, a process flow diagram should be developed to evaluate each process step, from receiving raw materials to shipping the product, to ensure that significant product safety hazards have not been overlooked or underestimated. The flow diagram should represent all process steps under the control of the company and may include steps prior to and after the plant's operations. The auditor should ask who or what group has the primary responsibility for each HACCP step. Typically, the personnel responsible for developing a flow diagram are members of the HACCP team. Often, the best results are achieved by including personnel from the engineering, maintenance, quality, and production departments. These employees' practical and technical knowledge of the process and equipment makes them valuable team members.

During initial development, the flow diagram should be very detailed. Each step in the process or movement of product through the manufacturing process should be noted. All steps from receiving to shipping must be identified so members of the HACCP team can use their combined knowledge to analyze potential product safety hazards. Biological, chemical, and physical hazards that are deemed significant and reasonably likely to occur, or that are inherent in the raw materials, must be reviewed for appropriate controls during the hazard analysis assessment. After hazard analysis assessment has been completed, the flow diagram may be simplified to make it easier to understand and to clearly represent the placement of the CCPs. In addition, the organization may want to develop a process schematic flow diagram. This flow diagram is useful for identifying how products, individuals, and utilities flow through the plant. It is a powerful tool for identifying areas where cross-contamination can take place.

For simplicity and ease of understanding, process flow diagrams usually are represented in block formation. The auditor should be flexible and accept any reasonable format for flow diagrams as long as the content is accurate and understandable. Both handwritten flow diagrams and computer-generated models are acceptable since the method used often depends on the resources available within the organization.

Verifying the Accuracy of the Process Flow Diagram

The process and responsibility for verifying the accuracy of a process flow diagram should be clearly stated in the organization's HACCP quality system procedures. The company may elect to have the HACCP team physically "walk through" the entire process from receiving to shipping to gain consensus as to whether the flow diagram clearly depicts the process. The company should have a valid reason for the manner in which it has chosen to construct the flow diagram. Where there are considerable amounts of raw materials, processing equipment such as receiving stations or conveyors, or inspection and testing prior to processing, the company may elect to break out receiving into a separate flow diagram for clarity and accuracy in hazard analysis review. This approach, if appropriate, allows for a thorough review of the company's existing product safety procedures, generally referred to as PRPs for HACCP.

Failing to include a process step in the HACCP plan will result in an inaccurate representation of the process and could have disastrous consequences. The omission of a processing step could mean that the step was not subjected to the required scientific hazard analysis review for biological, chemical, and physical hazards. A HACCP auditor should review the flow diagrams during the plant audit and understand how they were constructed.

In regulated HACCP, process flow diagrams usually are signed and dated to serve as a record that they were officially reviewed. The HACCP auditor should review for accuracy all flow diagrams at the site being audited. The following question should be asked when reviewing process flow diagrams: What group or person is responsible for verifying the accuracy of the flow diagrams and how are the diagrams kept up to date?

The HACCP auditor should allow enough time to actually walk through the plant to sample the accuracy and content of the flow diagrams. If process steps are omitted or bundled (grouped together) on the flow diagram, the auditor should ask the HACCP team to explain why this choice was made. For example, most flow diagrams state that receiving is the first step. However, the auditor may note that several bulk raw materials are received at the facility in addition to palletized raw materials. The bulk systems may contain sieving or other control systems designed to protect product safety or integrity. The auditor should assess each situation individually by asking questions and requesting more documentation that proves that product safety systems are in effect.

Another potential area of weakness is the handling of rework product. Most processes generate some type of rework. If the auditor notes that rework is being placed back into the process flow, the question should be asked: How and through what activities are product safety controls being applied? In most cases, the application of the PRPs will allow for effective preventive controls, but the significance of omissions noted on a process flow diagram must be investigated and evaluated by the audit team.

The organization should have documented procedures that address how a change in manufacturing prompts a review or modification of the original process flow diagram. The quality system procedures should state what group is responsible for reacting to process changes that could affect the HACCP product safety systems and explain how that information should be communicated throughout the company. At a minimum, the process flow diagrams should be reviewed annually to ensure that any changes in the manufacturing process have been reviewed for their impact on the HACCP product safety systems. Examples of process changes include equipment replacement, equipment additions, line relocation, and significant equipment modifications. The HACCP auditor should ask about changes to the manufacturing process and look for evidence that the process flow diagram is still technically accurate. Figure 4.1 is a verified process flow diagram with the CCPs noted.

ESTABLISHING THE PREREQUISITE PROGRAM

An important part of the establishment of a successful HACCP system is the prior or simultaneous implementation of a product safety PRP. Elements of a PRP are the building blocks or foundation of the "house of product safety" (see Figure 4.2). The NACMCF specifies in its guidelines for application of HACCP principles that a food HACCP system should be built on a solid foundation of PRPs.[2] Prerequisites are procedures, including good manufacturing practices (GMPs), that address adequate and sufficient operational conditions to protect public health. These procedures include personnel hygiene practices; employee training; cleaning and sanitation procedures; product recall programs; design, operation, and maintenance of equipment, grounds, and facilities; water safety; and handling of product throughout manufacturing and distribution.

An effective HACCP system cannot be built without the underpinning PRPs. PRPs typically are not part of a HACCP plan, and items covered in prerequisites rarely are designated as CCPs. This concept has been well defined when applying HACCP in the food processing industry. The primary difference between CCPs and prerequisite controls is that prerequisites ensure that food products are wholesome and do not contain objectionable contaminants, whereas CCPs are established solely for the purpose of controlling significant life- or health-threatening food hazards. PRPs address these types of food hazards only in instances where the hazard analyses for ingredients/raw materials and for process steps indicate that such a hazard has a low likelihood of occurrence. For example, even though broken glass from overhead light fixtures can be a significant food hazard, glass control and shielding of glass in overhead lighting usually is designated as a PRP. This is because typically there is a very low likelihood or frequency of breakage incidents in food plants. Thus, CCPs address food safety only, while prerequisites overlap into product quality and may involve other types of controls, such as quality or control points and operational steps. Finally, since CCPs apply to food hazards at specific points or steps in production or the process flow, they are specific to individual products and production lines. Prerequisites such as sanitizing and employee hand washing typically are implemented across an entire facility.

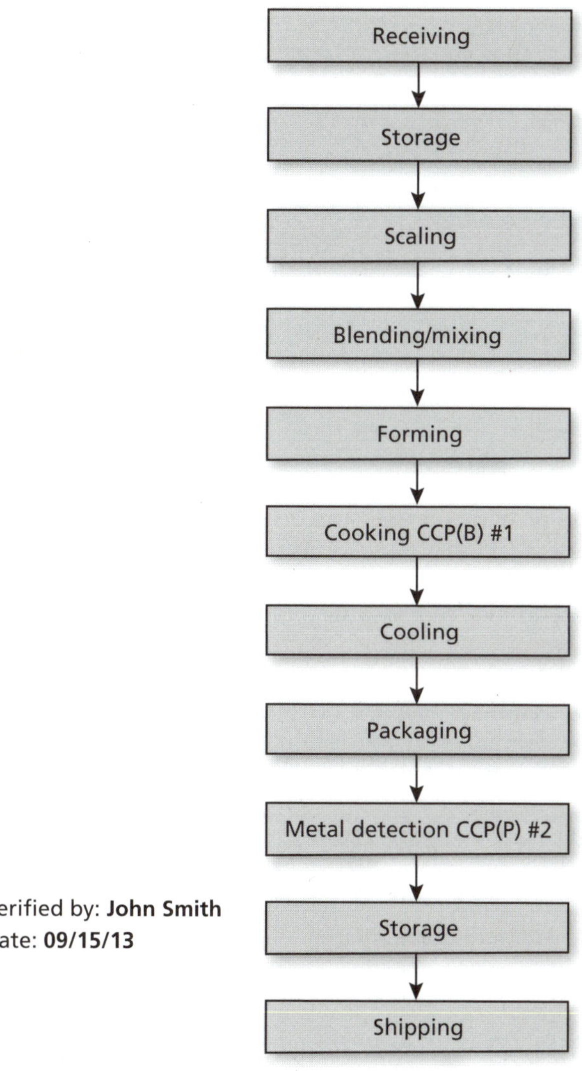

Figure 4.1 Verified process flow diagram.

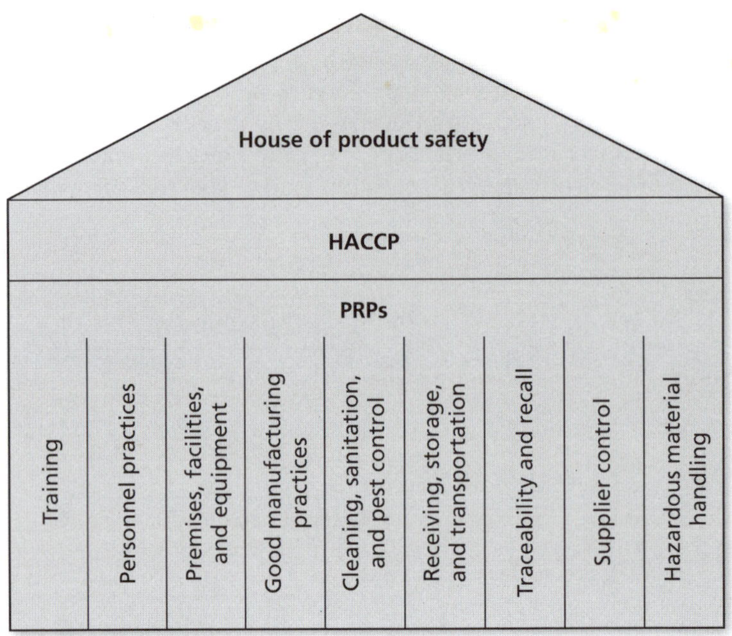

Figure 4.2 House of product safety.

Food hazards are biological, chemical, or physical contaminants that could cause illness or injury if ingested, including listeria, aflatoxin, glass, and metal. Contaminants that are noninjurious but are objectionable to the consumer are not hazards. Examples include burned product, hair, yeast, non-mycotoxin-forming molds, and food-grade lubricants.

Many ways exist to describe and categorize product safety PRPs depending on the regulatory perspective and industry sector. In some industry segments, a certain prerequisite may be of minor importance, while in others it may be essential and could even be designated as a CCP. Chapter 2 includes a more in-depth discussion of specific food safety PRPs.

NOTES

1. Donald A. Corlett Jr., *HACCP User's Manual* (Gaithersburg, MD: Aspen, 1998): 23.
2. National Advisory Committee on Microbiological Criteria for Foods (NACMCF), *Hazard Analysis and Critical Control Point Principles and Application Guidelines* (Washington, DC: U.S. Food and Drug Administration, August 14, 1997).

Sign & Date SSOP – on site implementation quarterly

Part II

Principles of HACCP

Chapter 5

Principle #1—Conduct Hazard Analysis

PURPOSE OF HAZARD ANALYSIS

Once the five preliminary tasks in the development of a HACCP plan have been completed, the HACCP team should undertake the first principle of HACCP: conducting a hazard analysis. Considered by many to be the foundation of a HACCP plan, the hazard analysis attempts to identify all potential hazards of a product, their sources, and the probability of their occurrence. Only then can appropriate control measures (product factors or processes that reduce or eliminate the hazard potential) be employed.

TYPES OF HAZARDS

A hazard is anything that could cause harm to a consumer using the product. All hazards must be identified scientifically so their potential risks can be assessed. Could a particular food or medical device hazard cause an illness, an allergic reaction, or a physical injury? If so, what can be done to prevent or minimize the possibility of the risk occurring?

Food Hazards

The HACCP system for food processing identifies three types of hazards that can occur in food products. They are classified by source. These hazards can be microbiological, chemical, or physical. In this book, allergens are treated as a subset of chemical hazards. Radiological hazards, discussed as a physical hazard in this book, are often considered chemical hazards in food protection. It is not as important to classify this hazard but make certain it is considered.

Microbiological Hazards

Pathogens and microbial toxins such as those listed in Table A.1 in Appendix A are significant hazards in many foods. Some individual ingredients and/or finished products have the potential to contain pathogens or allow development of microbial toxins that can cause mild to severe illness and even death. Additionally, permanent, lifelong debilitations can result from microbiological hazards.

There are two types of pathogenic microorganisms: non-spore-forming and spore-forming. A non-spore-forming pathogen is a foodborne microorganism recognized as a public health hazard that can cause illness or death in humans. Non-spore-forming pathogens include viruses, parasites, and bacteria. A spore-forming pathogen, on the other hand, is an organism capable of producing chemical- or heat-resistant spores. Upon outgrowth of the spores, the vegetative cells may produce toxins of public health significance that can cause illness or death in humans.

Chemical Hazards

As shown in Table A.3 in Appendix A, chemical contaminants in food may be naturally occurring or may be added during the processing of food. High levels of harmful chemicals have been associated with acute cases of foodborne illnesses, while lower levels can be responsible for chronic illness.

Potential chemical hazards include mycotoxins, antibiotics, pesticides, and sulfites. In most cases, due to the low likelihood of occurrence and/or the nature of the hazard, PRPs are the best method of control. However, in certain instances a chemical hazard may be recognized as a CCP and controlled as such. For example, many chemical substances, and nearly every food or food ingredient, can potentially cause an adverse reaction in at least one individual. However, a small group of substances called allergens are known to cause severe, life-threatening reactions that affect larger population groups. A substance is classified as an allergen if one or more of the following criteria exist: documented cases (published in scientific or medical journals) of severe, life-threatening reactions; several independently reported cases of these reactions; or clear scientific evidence or validation of the reaction by an expert experienced in the area of allergic reactions.

The Codex Alimentarius Commission recognizes that many life-threatening reactions to food substances can be avoided if the allergic person is aware that the substance is present. Latex in gloves, peanuts in a cookie, and penicillin in a prescription of antibiotics are all potentially allergenic substances that can be avoided if their presence is communicated through proper product labeling.

For guidance on which allergenic materials meet the aforementioned criteria, consult with authorities on the topic. The list of materials will be different depending on the industry and how the material will come in contact with or be ingested by the consumer.

Physical Hazards

In general, physical hazards are any objects or materials that: (1) are part of the product but are meant to be removed (such as bones in meat); or (2) are not designed to be part of the product but may be inadvertently introduced into the product during the production process (such as pieces of glass, metal, hard plastic, and so on).

Extraneous matter does not usually present a significant risk of a severe adverse health effect. Minor or moderate injuries such as those described in Table A.4 in Appendix A are more common when physical hazards are encountered.

While control of extraneous matter inherent in product raw materials (for example, bones, cherry pits, nut shells) is important for quality, the risks associated with these materials generally are less severe. Detection/removal devices for these objects are not necessarily managed as CCPs. PRPs such as supplier selection and approval and preventive maintenance are usually the best controls for the elimination or reduction of extraneous matter in products. However, in some cases, the characteristics (size, shape, and type) of the extraneous matter may potentially cause serious harm such as internal injury or choking. On that basis, some physical hazards must be managed as CCPs and controlled through appropriate measures such as detection/removal devices.

CONDUCTING A HAZARD ANALYSIS

Hazard analysis consists of two steps: hazard identification and hazard evaluation. Hazard identification involves analyzing each raw material, the production process, and consumer use. It also includes identifying appropriate control measures to reduce or eliminate potential hazards. Hazard evaluation is the process of reviewing each hazard that is identified to determine the severity of the health risk to the consumer and the probability of occurrence.

There is no one way in which to complete a hazard analysis. It is imperative that a cross-functional team with appropriate technical experts is involved in the hazard analysis, and that the evaluation is performed on the actual product and process. As the team progresses through the hazard analysis, it will identify and document hazards as well as control measures. Figure 5.1 gives examples of questions to be considered by the HACCP team when conducting a hazard analysis for a food product.

Hazard Identification

The HACCP team must examine all factors that have an impact on the safety of the final product, as well as characteristics of the product at each stage of production through distribution and consumer use. The process of hazard identification is actually quite easy once the preliminary tasks discussed in Chapter 2 have been performed. A good HACCP team made up of the right technical experts will be able to identify the potential hazards for most materials very quickly.

The preliminary tasks for the development of a HACCP plan include describing the product and its distribution as well as describing its intended use and consumers. This information can be used to evaluate any intrinsic factors of the product that could cause or prevent a risk to the consumer. For example, the inherent characteristics of two types of beverages—carbonated soft drinks and milk—can be compared to show how the risk of hazards differs due to the chemical makeup of these products.

Milk is from an animal; it is a neutral-pH product, and it contains protein and sugar (lactose). These factors cause it to be easily contaminated by pathogenic microorganisms, as is evidenced by the history of the product. In addition, the environment of the milk provides an excellent growth medium for microorganisms. Without treatment to kill the pathogenic microorganisms, there is an increased probability that consumers who drink the milk will get a foodborne disease.

Ingredients

- Does the food contain any sensitive ingredients that may present microbiological hazards (for example, salmonella, *Staphylococcus aureus*), chemical hazards (for example, aflatoxin, antibiotic, or pesticide residues), or physical hazards (for example, stones, glass, metal)?
- Are potable water, ice, and steam used in formulating or in handling the food?
- What are the sources of the ingredients (for example, geographical region, specific supplier)?

Intrinsic Factors

- What hazards may result if the food composition is not controlled?
- Does the food permit survival or multiplication of pathogens and/or toxin formation in the food during processing?
- Will the food permit survival or multiplication of pathogens and/or toxin formation during subsequent steps in the food chain?
- Are there other similar products in the marketplace? What has been the safety record for these products? What hazards have been associated with the products?

Procedures Used for Processing

- Does the process include a controllable processing step that destroys pathogens? If so, which pathogens? Consider both vegetative cells and spores.
- If the product is subject to recontamination between processing (for example, cooking, pasteurizing) and packaging, which biological, chemical, or physical hazards are likely to occur?

Microbial Content of the Food

- What is the normal microbial content of the food?
- Does the food product change in such a way as to allow for the growth of microorganisms?
- Does the microbial population change during the normal time the food is stored prior to consumption?
- Does the subsequent change in microbial population alter the safety of the food?
- Do the answers to the above questions indicate a high likelihood of certain biological hazards?

Facility Design

- Does the layout of the facility provide an adequate separation of raw materials from ready-to-eat (RTE) foods if this is important to food safety? If not, what hazards should be considered as possible contaminants of the RTE products?
- Is positive air pressure maintained in product packaging areas? Is this essential for product safety?
- Is the traffic pattern for people and moving equipment a significant source of contamination?

Equipment Design and Use

- Does the equipment provide the time/temperature control necessary for safe food?
- Is the equipment properly sized for the volume of food that will be processed?
- Can the equipment be sufficiently controlled so that the variation in performance will be within the tolerances required to produce a safe food?
- Is the equipment reliable or is it prone to frequent breakdowns?
- Is the equipment designed so that it can be easily cleaned and sanitized?
- Is there a chance for product contamination with hazardous substances such as glass?
- What product safety devices (for example, metal detectors, magnets, sifters, filters, screens, thermometers, bone removal devices, dud detectors) are used to enhance consumer safety? How are these safety devices calibrated and maintained?
- To what degree will normal equipment wear affect the likely occurrence of a physical hazard (such as metal) in the product?
- Are allergen protocols needed in using equipment for different products?

Figure 5.1 Examples of questions to be considered in hazard analysis of food products.
Source: National Advisory Committee on Microbiological Criteria for Foods (NACMCF), *Guidelines for Application of HACCP Principles,* Appendix C (Washington, DC: U.S. Food and Drug Administration, August 14, 1997).

Packaging
- Does the method of packaging affect the multiplication of microbial pathogens and/or the formation of toxins?
- Is the package clearly labeled "Keep Refrigerated" if this is required for safety?
- Does the package include instructions for the safe handling and preparation of the food by the end user?
- Is the packaging material resistant to damage, thereby preventing the entrance of microbial contamination?
- Are tamper-evident packaging features used?
- Is each package and case legibly and accurately coded?
- Does each package contain the proper label?
- Are potential allergens in the ingredients included in the list of ingredients on the label?

Sanitation
- Can sanitation have an impact on the safety of the food that is being processed?
- Can the facility and equipment be easily cleaned and sanitized to permit the safe handling of food?
- Is it possible to provide sanitary conditions consistently and adequately to assure safe foods?

Employee Health, Hygiene, and Education
- Can employee health or personal hygiene practices impact on the safety of the food being processed?
- Do the employees understand the process and the factors they must control to assure the preparation of safe foods?
- Will the employees inform management of a problem that could impact on safety of food?

Conditions of Storage Between Packaging and the End User
- What is the likelihood that the food will be improperly stored at the wrong temperature?
- Would improper storage lead to a microbiologically unsafe food?

Intended Use
- Will the food be heated by the consumer?
- Will there likely be leftovers?

Intended Consumer
- Is the food intended for the general public?
- Is the food intended for consumption by a population with increased susceptibility to illness (such as infants, the aged, the infirm, immunocompromised individuals)?
- Is the food to be used for food service, institutional feeding, or in the home?

Figure 5.1 Examples of questions to be considered in hazard analysis of food products *(continued)*.

Carbonated soft drinks, on the other hand, generally are produced from refined chemicals or processed agricultural materials, highly purified water, and possibly sugars. The pH of the finished product is between 2 and 3. Therefore, limited opportunities exist for the product to become contaminated with pathogenic microorganisms, as supported by the product's history.

Milk has an inherent microbial risk that must be controlled; carbonated beverage products have an inherent microbial safety. Soft drinks, in fact, have a low probability of causing foodborne disease provided the product meets the specified design. Other intrinsic factors that are part of a product evaluation include processes that make a product consumable, such as the baking of a cake, the frozen storage of ice cream, the dry nature of a vitamin pill, and the filtering of ground coffee. These factors contribute to making the product consumable, and thus safer, but such factors are often beyond the manufacturer's control. While evaluating intrinsic factors and examining how they are handled throughout processing can eliminate or reduce the occurrence of many hazards, to assure the safety of a product the HACCP team also must examine how the consumer might use the product. The intrinsic factors of a product's raw materials, the process, and consumer use become significant parts of the hazard analysis.

Raw Materials

Each material used to make a finished product must be evaluated for its potential to present physical, chemical, or microbiological hazards. All materials that can be incorporated into the finished product or that can be put into a consumer's mouth must be evaluated. This includes product that has been reworked, recycled, and reclaimed as well as processing aids, packaging materials (including shipping and storage containers), subcomponents, and water and steam sources.

As each material is evaluated, the identification of the potential for the hazard is based on the safety history of the material—scientific and/or historical evidence of the presence of the hazard. A *sensitive raw material* is any material that is likely to contain pathogens or toxins and/or that allows the growth of any pathogen. The definition is sometimes expanded to include raw materials that are historically known to contain physical or chemical hazards. The HACCP team must agree on the operational definition being applied to "sensitive raw material" so everyone understands what is meant by the term.

After all raw materials for a product have been identified, including processing aids and additives, they must be described in detail and documented. What is the material? What are its intrinsic factors? Who is the supplier? What is its function in the finished product? How is it manufactured, stored, packaged, and distributed? All these factors will have an impact on the safety or risk potential that a material may bring to the finished product. These items should be described in enough detail so anyone picking up the documentation will have a clear understanding of exactly what the material is, even without physical samples of the material. This ensures that HACCP team members have access to all the information needed to provide an accurate assessment of the material during hazard analysis.

Key pieces of information to identify about each material are its physical state and how it is handled. Physical state is a description of the material. Is the material wet or dry? What are the pH, water activity (A_w), and types of acidulants? Does the product contain fermentable carbohydrates or preservatives? Does the product contain any antimicrobial additives? What is the size and shape of the material? Is it alcohol or water based? Is it processed or direct from the field (a raw commodity)? Handling refers to how the material is processed, packaged, stored, and distributed. Is the material heat-processed? Is it ground and sifted? Is it filtered? Is it shipped in tanker cars or bagged in low-density, polyethylene-lined, 45-pound bags? Is it stored refrigerated or at room temperature?

After the material has been defined, and the process by which it is made and handled is known, types of hazards that could be introduced into the finished product from the individual raw materials must be identified. It is not enough just to state that pathogenic microorganisms could be present in the material. The

HACCP team must specify which organisms could be present, then describe the severity of the health consequences if the identified hazard is not controlled.

The likelihood of occurrence of the potential hazard needs to be determined based on documented historical or scientific evidence. Similar materials can be considered when determining the evidence of risk. There may be little or no documentation available on a particular material because it has not been widely studied or used. However, a lack of scientific evidence does not mean that risks do not exist, especially when hazards have been associated with similar products already in use. Table 5.1 provides an example of documentation for a raw material hazard analysis.

One of the most difficult aspects of completing the raw material evaluation is understanding the supplier. Research and actual site inspections may be required to obtain a clear and concise review. Not all raw materials will require the same degree of evaluation. For instance, some raw commodities will undergo processing steps specifically designed to eliminate hazards. Suppliers of these materials will be subject to less scrutiny than suppliers of other materials because it is expected that unprocessed commodities may contain hazards. For this type of material, the degree of control (reduction of hazards) will often fall to the manufacturer of the finished goods. For foreign suppliers, an effective strategy must be implemented to ensure that the foreign raw materials are safe and meet appropriate federal regulations.

If the material is processed, however, the purchaser often expects that potential hazards have been eliminated or reduced to an acceptable level and that there is a low probability of occurrence. The control of the hazard will then be the responsibility of the manufacturer of that material. In these cases, verification that the supplier has implemented an effective product safety management system is critical to assuring the safety of the final finished goods.

ISO 22000 or other food safety certification can provide verification that the company has a structured, documented, and implemented quality management system. The certification alone, however, is not a guarantee of safe products. If the quality management system fails to identify key product safety activities, there is a potential for product safety risk.

A site audit is the best method for gaining a complete understanding of the quality system of each supplier. All manufacturing sites that supply materials should be audited before entering into a contract with the supplier and receiving any materials. It is important to be assured that a supplier has effective GMPs and product safety programs in place. Ideally, all suppliers also will be using HACCP as a key part of their product safety program.

The purchaser should research the product quality performance history of each supplier. Information on product recalls is part of the public record. This allows gathering of information on a product type or a company history by requesting the information on the recall through the Freedom of Information Act or by checking appropriate websites. The FDA publishes Enforcement Reports for a number of years. If a product type or a company has a record of multiple product recalls in the recent past, the purchaser needs to utilize this information when identifying potential hazards and the probability of the risks occurring. This recall information should be considered when completing the hazard analysis.

Table 5.1 Raw material hazard analysis documentation.

Ingredient	Description	Storage	Determine potential hazards associated with material	Assess severity of health consequences if potential is not properly controlled	Determine likelihood of occurrence of potential hazard if not properly controlled
Pasteurized liquid whole egg	Liquid whole eggs are composed of both the yolk and the white from chicken eggs. The eggs are washed, checked for quality, and mechanically cracked. The liquid egg is then pumped into a large tank through a 60 mesh filter and held at 45°F until enough material is accumulated for pasteurization. It is then pasteurized at 191°F for 10 seconds and rapidly cooled to 40°F. Whole liquid eggs are shipped in tanker trucks that are held at <42°F. Material is liquid, with a neutral pH. It contains fermentable carbohydrates and is an excellent growth medium for microorganisms.	After pasteurization, eggs must be held at no greater than 45°F for no greater than five days.	Microbiological—salmonella in finished product	*Salmonellosis* is a foodborne infection causing moderate to severe illness that can be caused by ingestion of only a few cells of salmonella.	Product is made with liquid eggs, which have been associated with past outbreaks of *salmonellosis*. Recent problems with *Salmonella serotype Enteritidis* in eggs cause increased concern. Probability of salmonella in raw eggs cannot be ruled out. If not effectively controlled, some consumers are likely to be exposed to salmonella from this food.
			Chemical—eggs are a known food allergen	Egg-allergic individuals can suffer health consequences from mild to severe. There have been documented cases of death caused from severe allergic reactions to eggs.	As an ingredient in the finished product, unless it is properly identified as in the ingredient statement, there is a probability of the product being consumed by a food-allergic individual.
			Physical—eggshells	There are no documented cases of severe injuries caused by eggshells. There are cases of minor mouth abrasions.	There is a low probability of any injury from eggshells, since the liquid eggs are filtered through a 60 mesh screen and the shells are brittle.

Source: Modification of National Advisory Committee on Microbiological Criteria for Foods (NACMCF), *Guidelines for Application of HACCP Principles,* Appendix D (Washington, DC: U.S. Food and Drug Administration, August 14, 1997).

It is not in the best interest of a company to buy any material without a complete understanding of the supplier's organization, including knowledge of its financial stability. It also is important to ensure that the organization can supply the materials needed in a timely manner while meeting all specifications.

The last step in the analysis of a supplier is understanding the supplier's contract manufacturer practices, if applicable. A common practice is to have a contract manufacturer make product. The purchaser must ensure the supplier understands that all approvals are contingent upon the approval of the manufacturing location that is to supply the material. If the supplier uses contract manufacturing, the purchaser should ensure the supplier has good quality and product safety systems for the approval and oversight of contract manufacturers. In the absence of such programs, the purchaser should retain the right to refuse material from a specific contract manufacturer.

Process Review

After all materials used to make the finished product have been evaluated and any potential hazards have been identified, the HACCP team should evaluate each step in the process (from receiving to consumer use) and identify all potential points in the process where hazards can be introduced. Aspects of the processing environment to consider include facility design, traffic flow, equipment design, and the function of a specific processing step. Process steps that are designed to reduce hazards also should be identified at this time.

A good, documented product safety system includes an understanding of all the steps involved in the manufacturing of a product. Some of the questions that must be addressed about each step of the process include: What is done and by whom? What training do they receive? What records are kept?

Consider, for example, the process of receiving dried whole eggs into a facility. The specification for the material states that a certificate of analysis (COA) must be received at the facility before or at the time of the shipment. The following paragraphs explain what is done during this step of the process and why.

Material arrives at the shipping and receiving dock of a facility, and the driver provides the shipping clerk with the bill of lading. The shipping clerk checks the raw material against specification requirements. The specification states that a COA is required for each lot of material. Microbiological testing to be done includes tests for salmonella, fat, and moisture. If the COA has not been received prior to delivery, then it must be included in the documentation accompanying the delivery.

It is important for the HACCP team to understand why the specification requires pathogen testing to be performed and why the supplier does the testing rather than the receiving facility. Of the required tests, two are indicators of product safety: the tests for salmonella and moisture level (if the moisture level is too high, the eggs could provide a media for microorganism growth). The supplier performs the tests to reduce the risk of the purchaser introducing and potentially using a contaminated material by assuring that the microbiological hazard has been controlled even before the product is brought into the purchaser's facility.

When the shipping clerk knows the importance of the COA, the clerk will understand exactly what to do. The shipping clerk should compare all packages of the material to the lot numbers indicated on the COA. If a COA is not on file for a lot of material, then the lot must be rejected. The shipping clerk's work instructions must clearly state the steps to be followed when a deviation is found, such as the absence of a COA for a lot of material.

If the delivery is found to be acceptable, the shipping clerk should file the information, including the COA, and make the material available to the operation for processing. This one step ensures that the microbiological hazard of the dried whole eggs is being controlled at the supplier by providing appropriate documentation via the COA.

This process of checking prevents the use of an untested and potentially contaminated material in the manufacture of finished product. The HACCP team must determine if this step of checking a shipment of products for the proper COA is or is not a CCP. This decision is based first on identifying the processing step and potential hazards and then determining the specific mechanisms for controlling the hazards. In addition, if this step is selected as a CCP, the HACCP team will need to validate the process of checking the COA.

This same process needs to be completed for every step in the manufacture, distribution, and use of a product. Each step must be regarded as potentially introducing and/or eliminating a hazard. Figure 5.1 and Table 5.1 can provide guidance on the types of questions that must be asked when completing this step of the hazard analysis. During a HACCP audit, the auditor should ask: Who is doing what, with what? When is it being done, where is it being done, why is it done, how is it done?

Someone who actually does the task should be part of the HACCP team for this step of the analysis. Ideally, the discussion should include an observation of the process to verify that the tasks are being performed in accordance with the documentation.

Allergen Review

The allergen review identifies the risk of contamination by unlabeled allergens through equipment cross-contamination. It is possible for an allergenic material to be accidentally incorporated into a product if the product is made on the same production equipment as another product that contains the allergenic material. Therefore, the HACCP team needs to analyze what is actually produced on a specific production line.

The first step is to establish whether other products are produced on the same production line and to determine if any of these products contain an allergenic material. Next, the following question should be asked: Does the product that is covered by this specific HACCP plan contain the same allergenic materials? If the answer is "yes," then confirming that the product labeling clearly identifies the allergen is the control measure. If the answer is "no," the product in question does not have the same allergen profile as other products made on the same production equipment. This increases the potential for the introduction into the product of an allergenic material that is not identified on the label. The HACCP team needs to ensure that an appropriate control measure for the allergens is in place. The HACCP team may need to determine the effectiveness of the cleaning and sanitizing program in reducing the potential allergen contamination to an acceptable level. Other control measures include labeling the end product to reflect potential allergens, controlling production schedules to ensure products that contain allergens are manufactured after products that do not contain the allergens, properly storing product or ingredients that contain allergens, and providing allergen warning labels on ingredient packages and ingredient formulation sheets.

Consumer Use and Identification of Control Measures

When a hazard is identified, the control measure(s) for it must be identified. The HACCP team must ask the question: How is the consumer protected from this hazard? Not all control measures will be within a manufacturer's control; often it will be the consumer's responsibility to control the hazard. In such cases, it is the manufacturer's duty to inform the consumer of potential risks by including instructions for safe product use. Dosage restrictions on medications, such as "Do not exceed six tablets in 24 hours," or directions, such as "Keep refrigerated," on shell eggs are examples of statements provided by the manufacturer on product packaging to help ensure products are stored properly and used in the recommended amounts. Sterile medical devices commonly have a statement declaring that the product must be used before a certain date or that the contents are sterile only if the package has not been opened.

All control measures must have an identified scientific basis for being an effective means of control. For example, certain time/temperature applications have been proven to kill microorganisms. A dry material, or material with low water activity (A_w), will not promote the growth of microorganisms.

If it is found that a processing step or a material actually introduces a hazard into the product, the hazard should be designed out of the process when possible. For example, if eggs are added to a product solely for flavor, can the product be reformulated using nonmicrobiologically sensitive material? This replacement would eliminate the potential for the introduction of salmonella into the finished product.

In those cases where a hazard is identified and no known control measure is in place, it will be necessary to design a control measure, with a scientific basis, into the process or product.

Hazard Evaluation

After hazards and their appropriate control measures have been identified, each hazard should be evaluated for severity and probability of risk. This needs to be done before establishing whether the control measure is a CCP or is part of a PRP.

Severity of Risk

The severity of a risk is a difficult thing to assess. Severity often is judged based on a scale of high, moderate, and low, with high being life-threatening reactions or those causing irreversible organ damage or failure and low being minor reactions or reversible and treatable medical conditions. The most common reaction to a hazard may be low to moderate; however, in certain individuals or population groups (the aged, infirm, immunocompromised, or infants) the health consequences may be life-threatening.

Obviously, the most severe risk of any hazard is death. No ethical company would knowingly create a product that would harm or kill anyone. However, historical evidence exists of products that have hurt and killed people. In some cases, the cause was ignorance of the risk. In other cases, changes in the environment brought the hazard to light. Either way, the manufacturer is obligated to eliminate or reduce the risk of harm.

A number of factors must be considered in assessing the potential health and safety risk to the consumer if a hazard is not controlled. The HACCP system is not as effective for controlling hazards with consequences that tend to manifest themselves over the long term and cannot be directly correlated with the ingestion or use of a specific product. HACCP is most effective with health consequences that are immediate and that can be traced to the actual product ingested or used. For example, compare the potential for liver cancer caused by the consumption of mycotoxins in grain versus the potential for salmonella foodborne disease from eating a raw egg. It is easy to correlate the eating of a raw egg to a case of foodborne disease that manifests itself within 24 hours of consumption. On the other hand, it is much more difficult to identify whether a specific lot of grain eaten 30 years prior could have had an elevated mycotoxin level that may have caused liver cancer. Many times the person had been exposed to other chemicals that can cause liver cancer.

When considering the severity of a hazard, the following questions need to be asked: What potential customers have the highest severity of risk for the hazard? What are the health consequences (mild to severe) if exposed to the hazard? What is the potential duration of the illness or injury? If the HACCP team does not know the answers to these questions, it should seek professional advice. State and federal regulatory agencies often publish epidemiological summaries or morbidity/mortality reports that establish the hazard profiles of different products. The severity of a risk is not lower just because the HACCP team is unfamiliar with the hazardous health consequences.

Probability (or Likelihood) of Risk

The final and most difficult part of the hazard analysis is the assessment of the probability of risk. While difficult, the decisions made on probability can mean the difference between a focused, effective HACCP system and one that is ineffective. The latter occurs when the HACCP system has too many CCPs. This makes the system difficult to manage and overly burdensome to the organization.

One factor to consider when trying to establish the probability of a hazard is product history. Has the hazard previously been found in this product/material? If possible, identify the source of contamination and determine whether patterns exist or whether the problem appears to occur in a random manner. The frequency of occurrence also should be identified. The HACCP team should determine how many times the hazard previously occurred.

Another factor is the control measure. Is the control measure highly dependable and in statistical control within very specific limits, such as heat treatment or pH? This will affect the probability of occurrence. The greater the predictability of the action, the better the understanding of the probability.

If the HACCP plan is being designed for a new product or a product with no clear safety history, the HACCP team should look at the safety performance of other similar products. Have hazards been found in similar products? Products undergoing similar manufacturing processes (for example, dry mix) and products containing common raw materials (such as egg-containing products) should be examined.

One way to determine whether the hazard is common and the risk is severe is to look at the domestic and international regulations and risk assessment documents for the industry. In general, if a regulatory body has addressed the hazard and prescribed a specific control measure, the process was accomplished using scientific evidence. Another source of information is product safety actions taken by companies or regulatory agencies. A product recall is validation of the probability of occurrence of a specific hazard, especially if it happens more than once and to more than one company. The FDA regularly publishes notices of food and medical device recalls. A review of those recalls can be used to establish a probability of occurrence. Sterility problems always underlie a large number of the total recalls of medical devices. While the methods for sterility assurance have high confidence values, recalls for inadequate sterilization based on related problems, including packaging, are common.

Many industry trade associations and regulatory bodies are developing model HACCP plans to assist companies. Since many smaller organizations do not have the internal resources to perform a HACCP analysis, model plans provide a good starting point for information on types of hazards that can be expected in a product. There is, however, a significant risk to the direct adoption of any model HACCP plan. These plans are very conservative. HACCP plans should be specific to a product and production line. HACCP plans are one-of-a-kind plans and must be reviewed with every change in raw material or process. One minor change in the product or process may introduce a significant uncontrolled hazard. Model HACCP plans are developed to be generic; they do not and cannot take into account the specifics that make up any finished product. It is impossible for a model plan to identify hazards that are the result of a supplier's history, allergen cross-contamination, or even the intrinsic factors for any specific finished product.

HACCP is designed to protect against hazards that are reasonably likely to occur. It is not designed to protect against every potential, random, accidental, or intentional occurrence. What is the probability of an employee's glasses accidentally falling into a mixer and going unnoticed? It has happened. In all the millions of packages of products over decades of time with thousands of different employees, how many times has this occurred? It would be very difficult, if not impossible, to find any record of such an event. HACCP is about due diligence. It is the proactive identification of hazards far in advance of incidents of injury or illness occurring.

Hazard evaluation includes probability and possibility. It is possible for anything to happen. But is it probable? It is possible to get struck by a meteor and lightning at the same time, but it is not very probable. The HACCP team and the HACCP auditor should ask the following question: Is this hazard reasonably likely to occur under these given conditions? If it is unreasonable to expect the hazard to occur, then it is a mere possibility.

The issue of probability and HACCP is a highly debated topic. There is no clear method for determining with absolute certainty all the hazards that should be managed in the HACCP plan. Unless there is regulatory guidance, the final identification of hazards and their potential risks is a management decision. The process of analyzing hazards is constantly being challenged as new processes and products are developed, and as new information is obtained and new hazards become known. The key is doing the best job possible each and every time a HACCP plan is developed or verified using the best resources available.

DOCUMENTATION AND ONGOING EFFORTS

Documentation of the hazard analysis must be complete, clear, and made readily available to the organization and the HACCP auditor. The complete hazard analysis, with all supporting documents—including, but not limited to, references, audit reports, and scientific evidence—should be kept on file in one central location. An electronic document control system provides a way to store and control the HACCP-related documents while providing easy access to all employees who need to read and review the documents. When an electronic system is used, all printed documents should be labeled as "uncontrolled," and it should be noted that the records may be out of date. A statement should be included to the effect that the electronic record is the only controlled document.

On a local basis, the hazard analysis documentation for each material and processing step evaluation should be available and be part of the final HACCP plan. The HACCP plan also should list all supporting documentation and identify the official file location.

The best way to do a hazard analysis is by using the three actuals: the analysis must be done on the actual product, at the actual production location, with the actual people who know the product and the potential hazards. The hazard analysis is based on facts, not assumptions and conclusions like "we've never had a problem." The hazard analysis requires research and technical knowledge about many different topics. The proper HACCP team with the right support and information is critical to the accurate identification of the potential hazards of a product. This identification is the foundation for an effective HACCP plan and the protection of consumers.

Chapter 6

Principle #2—Determine Critical Control Points

DISTINGUISHING BETWEEN CRITICAL CONTROL POINTS AND CONTROL POINTS

A *critical control point* (CCP) is defined as "a step at which control can be applied and which is essential to prevent or eliminate a food safety hazard or reduce it to an acceptable level."[1] The medical device industry uses the term *essential control point* (ECP) instead of CCP. An ECP is "a point, step, or procedure at which control can be applied and which is essential to prevent, eliminate, or reduce a hazard to an acceptable level."[2] Controlling factors or variables at CCPs is described as implementing control measures. Control measures, then, describe actions and activities taken at the CCP to prevent, eliminate, or reduce the identified hazard. Every significant hazard must have a control measure to reduce the likelihood of its occurrence. The control measures are dependent on the reliability of the food safety control system.

As "a point, step, or procedure" in the production process, a CCP does not focus on the supporting manufacturing infrastructure, such as sanitation, equipment maintenance, pest control, personnel programs, transportation and storage requirements, premise maintenance, and recall and traceability requirements. As will be discussed in Chapter 2, product safety issues pertaining to these and similar areas must be controlled through prerequisite programs (PRPs). A detailed set of PRPs outlines how product safety will be assured and simplifies identification of CCPs by focusing on process steps rather than plant infrastructure.

A CCP differs from a control point. A control point is "any step at which biological, physical, or chemical factors can be controlled."[3] As such, most control points usually are related to quality, production, or PRP issues. A control point normally is not associated with product safety unless the control point supports a CCP. For instance, a dry ingredient mix facility may place screens, magnets, and a metal detector in the production line to prevent metal contamination of the finished product. The screens and magnets are control points; only the final point of control, the metal detector, is a CCP. ISO 22000 uses the term *operational prerequisite program* (OPRP) instead of *control point*. ISO 22000 uses the following definition for OPRP: An OPRP is "PRP identified by the hazard analysis as essential in order to control the likelihood of introducing food safety hazards to and/or the contamination or proliferation of food safety hazards in the product(s) or in the processing environment."

To differentiate between a control point and a CCP, ask two questions. First: If I lose control of this step, is there a succeeding step (for example, a kill step, a chemical wash, a freezing step) that could effectively control the hazard? If the answer to this question is "yes," then the step is probably a control point. If the answer is "no," ask the second question: If I lose control of this step, could the product cause serious illness or injury?[4] If the answer is "yes," the step is probably a CCP.

COMMON SOURCES OF CRITICAL CONTROL POINTS

CCPs often are found in the areas of raw materials, ingredient receiving and handling, processing, packaging, and distribution.

Raw Materials

Product contamination by microbiological, chemical, and physical hazards—such as pathogens, pesticides, herbicides, antibiotics, naturally occurring toxins, and metal fragments—may have its source in raw materials. When a processor has control measures in place to prevent contaminated raw materials from entering the plant, these materials and the act of their receipt can be CCPs. This is especially true if no step exists in the process to eliminate or reduce the hazard—for example, a thermal processing step to eliminate a microbiological hazard. If a significant hazard may be associated with a raw material, then a supplier quality assurance program should be in place to control the hazard to the best of the supplier's ability. See Chapter 19 about using good agriculture practices (GAPs) to control the safety of raw agricultural products.

A raw material decision tree can be used to determine if an incoming raw material might be considered a CCP.[5] A raw material decision tree can be used to answer the following questions:

Question #1: Is a significant hazard associated with this raw material? This first question can be answered by the hazard analysis. If identified hazards are sufficiently severe and likely enough to occur that they could harm someone, then the answer to this question is "yes"; proceed to question #2. If the answer is "no," then ask question #1 of the next raw material.

Question #2: Will this hazard be processed out of the product? If no way exists to reduce a hazard to an acceptable level or eliminate it during processing, then this hazard probably occurs in the finished product. Options to control the hazard include adding a process step to reduce or eliminate the hazard. Another method is to ensure that customers know and understand that a hazard exists in the product and use a CCP to control the hazard prior to consumption of the product. A third way is to ensure that the supplier has control programs for the raw material so the hazard can be controlled before the raw product reaches the production plant.

Other less accurate methods of control are "hold and test" programs that require acceptable test results before a raw material can be used. Usually, COAs or "hold and test" methods are not considered acceptable means of preventing or eliminating hazards at the raw material level. It is unlikely that these methods will detect minute levels of contamination. In cases where they are the only possible methods of control, sampling and testing methods must be stable and capable of ensuring the accuracy and reliability of the results.

Question #3: Is there a cross-contamination risk to the facility or to other products that will not be controlled? If the answer to this question is "no," proceed to the next raw material. If the answer is "yes," then process steps or PRPs may be needed to eliminate or reduce this risk.

Ingredient Receiving and Handling

If an incoming raw material contains biological, chemical, or physical hazards, the manner in which it is received, handled, or stored might be a CCP. For instance, the improper storage of some dry ingredients can result in aflatoxin production. If control measures are not in place under the PRPs to reduce or eliminate this hazard, then these steps can be considered CCPs. Control measures for this type of CCP include sifters, magnets, and temperature and humidity control.

Processing

Process steps are commonly identified as CCPs. Examples of these steps include rework, cooking, chilling, and formulation control.

Rework

Rework and salvage processes may be CCPs, particularly if any of the products contain allergenic ingredients and a risk of cross-contamination with other products is possible. Control measures include production scheduling, product handling, sanitation, and mixing rework into identical products.

Chilling

Cooling or chilling may be a CCP. Bacteria spores could germinate or grow during the cooling or chilling process and become a serious health hazard. Therefore, both time and temperature variables can be CCPs if bacterial spores have not been destroyed with a cooking step, or if the growth of vegetative cells such as *Staphylococcus aureus* has not been prevented to preclude toxin formation.

Formulation Control

Formulation of the product may be a CCP. During formulation, ingredients can affect the product's ability to support microbial growth, cause allergic reactions, or adversely affect consumer health if maximum allowable limits are exceeded. Variables include ingredient proportions such as weights and volumes, pH, A_w, ingredient concentrations, ingredient inventory monitoring before and after the batch is mixed to ensure that the correct amounts of sensitive ingredients are used, adequate agitation or mixing times to ensure a homogeneous mix, and verification testing of the finished mix to ensure the correct usage of certain key ingredients. Thus, mixing of the product may be a CCP.

Packaging

Packaging is a step in the production process that should be considered during CCP determination. During the packaging step, a number of factors may be considered CCPs. For instance, the integrity of the package seal may be considered a CCP. Other activities at the packaging level that might be considered CCPs include the detection of metal or other foreign material and the presence of a proper vacuum or proper gas mixture in modified-atmosphere packaged products.

Ensuring that package ingredient declarations are correct can be a CCP in cases where ingredients may cause allergic reactions or have controlled regulatory health limits. Issues such as correct coding for traceability are usually considered part of a recall and traceability PRP.

Distribution

Time, temperature, and humidity might need to be controlled during the storage and transportation of a product. A comprehensive PRP for transportation and storage might be adequate to control the safety of the product. However, in some cases these variables are critical to the safety of the product. In those instances, a CCP might be identified at the storage and/or transportation steps. When in doubt, ask the question: If I lose control of this step, could the product cause serious illness or injury? If the answer is "yes," then the step may be a CCP.

Remember that CCPs are points, steps, or procedures under the manufacturer's control. If customers are responsible for transportation and storage, the CCP will be part of the customer's HACCP plan. Many factors that control hazards related to certain products are beyond the control of the manufacturing facility (for example, they may occur at the retail level, in the food service arena, or within homes). A medical device intended for hospital use is tested under certain fairly controlled environmental conditions. The same device may be used in a home environment or outdoors where the temperature is colder or warmer, or it may be used in the bathroom where the humidity is high. Those extremes may adversely affect the medical device. During the hazard analysis, the HACCP team should attempt to identify and document key health hazards, even ones that may be beyond the manufacturer's control. While it may be impossible to eliminate these hazards, steps may be taken to lessen their impact. For instance, temperature-indicating sensors may be incorporated into packaging to ensure that temperature abuse is evident, or label instructions may be added to identify key storage or handling requirements.

IDENTIFYING CRITICAL CONTROL POINTS

CCP decision trees are tools recommended for use during the CCP determination step. These trees provide the HACCP team with a systematic and logical approach to determining CCPs. Decision trees also provide a basis for documenting the reasons for selecting or rejecting a step as a CCP.

Considerations When Selecting a Decision Tree

The most widely used decision trees include those developed by the Codex Alimentarius Commission (1997 version), the NACMCF (1998 version), and the Canadian Food Inspection Agency (1995 version). These decision trees, as well as one commonly used in the medical device industry, are shown in Figures 6.1 through 6.5. Regardless of the decision tree used, the results obtained should be tested against the experience and knowledge of HACCP team members. Important considerations when using any decision tree include:

- Each process step identified in the flow diagram must be considered in sequence.
- At each step, the decision tree must be applied to all identified hazards.
- Use a decision tree to determine CCPs only after the hazard analysis has been completed and the significance of each hazard has been evaluated.
- A CCP may have more than one control parameter within its control measure. For instance, a pasteurization CCP could include both time and temperature variables.
- More than one hazard may be controlled by a specific control measure.[6]
- A specific hazard may need several control measures.
- The number of CCPs that may be identified is unlimited. However, the ideal HACCP plan will have a limited number of CCPs. Most of the hazard control will be accomplished using PRPs.

As shown in Figure 6.6, two questions should be answered before a decision tree is used to determine CCPs:

1. *Does this step in the process involve a hazard of sufficient likelihood of occurrence and severity to warrant its control?* This question is asked as part of the hazard analysis. If the answer is "yes," proceed to the next question. If the answer is "no," then this step is not a CCP. Ask this question of the next process step.

2. *Is this hazard fully controlled by a PRP?* Specific process steps, such as cooking or sifting, are never completely controlled by a PRP. PRPs need to be verified to ensure they are operating according to the PRP plan. Equipment prerequisite areas may include the calibration and preventive maintenance of an oven but not the actual cook time and temperature.

On the other hand, training and education programs on employee hand washing may be sufficient to control the hazard of "microbial contamination during employee handling." Another example is the ability of sanitation programs to fully control microbial or chemical hazards on equipment. If the answer to this question is "yes," then this is not a CCP; proceed to the next process step. If the answer is "no," proceed to the next question.

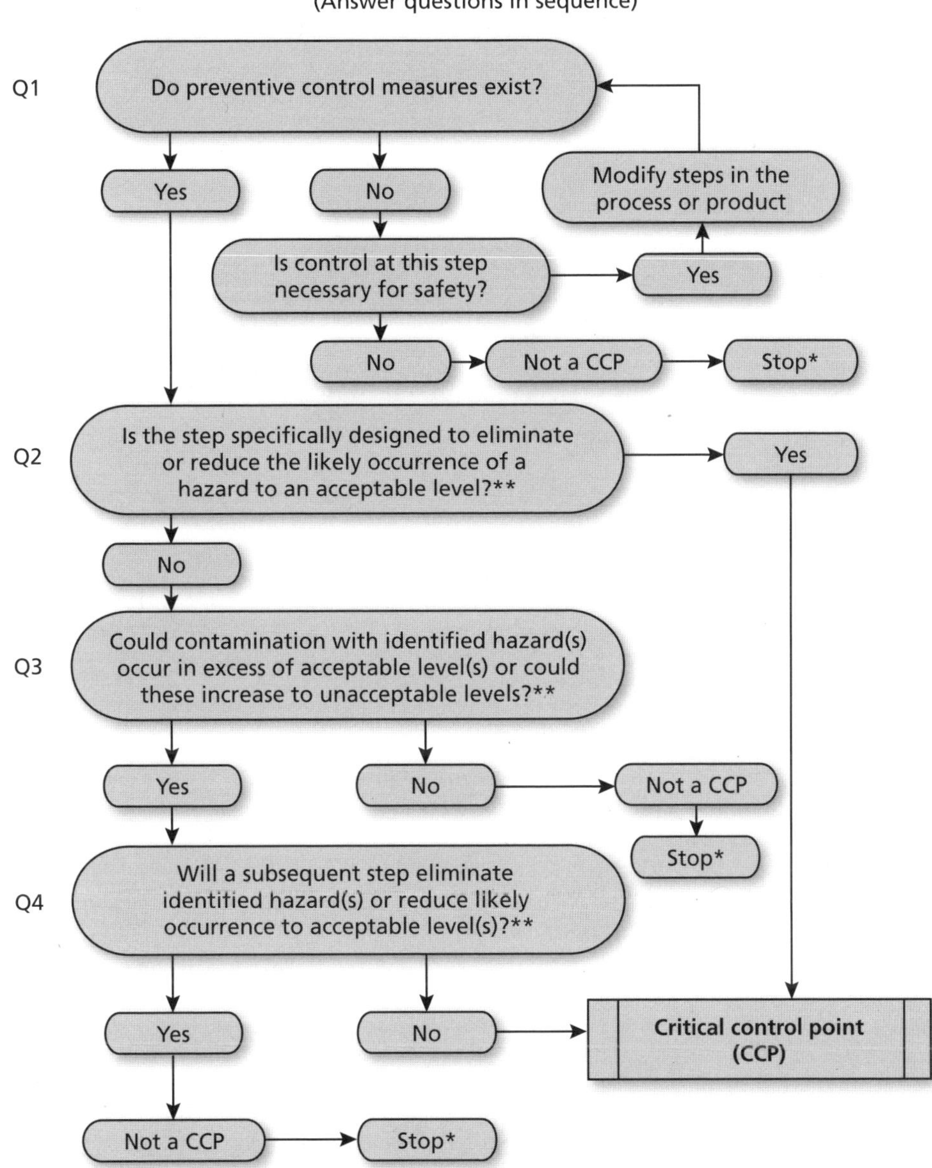

Figure 6.1 Codex Alimentarius Commission decision tree.

Source: Codex Alimentarius Commission, *General Principles of Food Hygiene,* CAS/RCP 1—1969 (Geneva, 2003).

Q1 Does this step involve a hazard of sufficient likelihood of occurrence and severity to warrant its control?

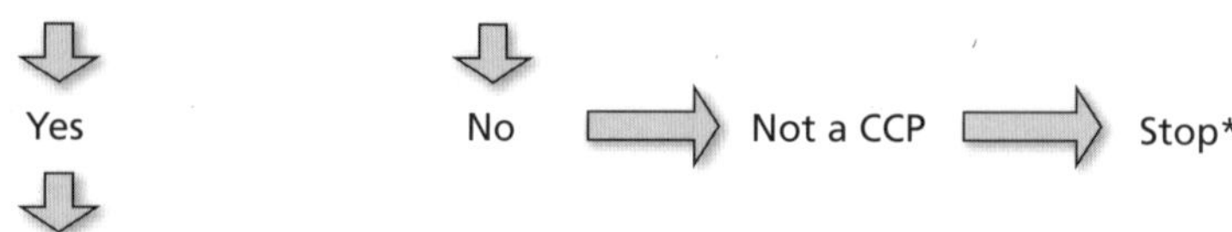

Q2 Does a control measure for the hazard exist at this step?

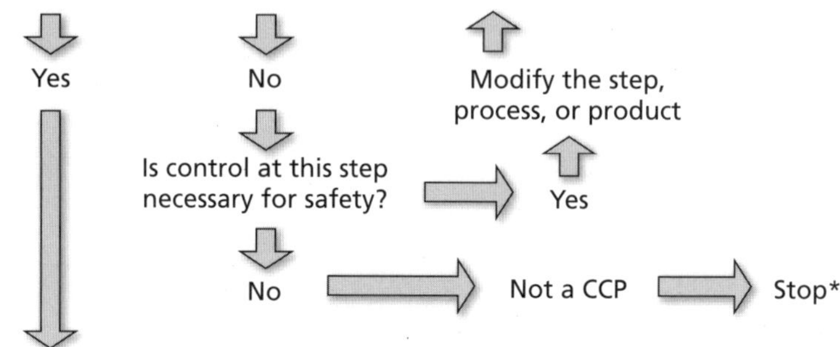

Q3 Is control at this step necessary to prevent, eliminate, or reduce the risk of the hazard to consumers?

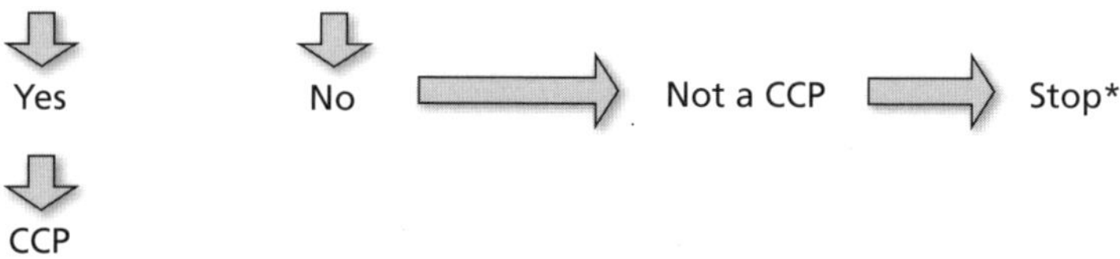

*Proceed to next step in the process.

Figure 6.2 NACMCF decision tree 1.

Source: National Advisory Committee on Microbiological Criteria for Foods (NACMCF), "Hazard Analysis and Critical Control Point Principles and Application Guidelines," *Journal of Food Protection* 61 (1998): 1246.

Using a Decision Tree

If the answers to both of the previous questions indicate that a process may be a CCP, then a decision tree may be used as a tool to determine CCPs and document the reason for their selection. The following questions are commonly asked on a decision tree, such as the one shown in Figure 6.7, to determine CCPs.

Question #1: Do control measures exist for the identified hazard? Question #1 asks whether the operator could use any control measure at this step or elsewhere in the process to control the identified hazard.

If the response to this question is "yes," clearly describe what measure(s) the operator could take to control the hazard, for example, "Yes—metal detector" or "Yes—cooking," then proceed to question #2.

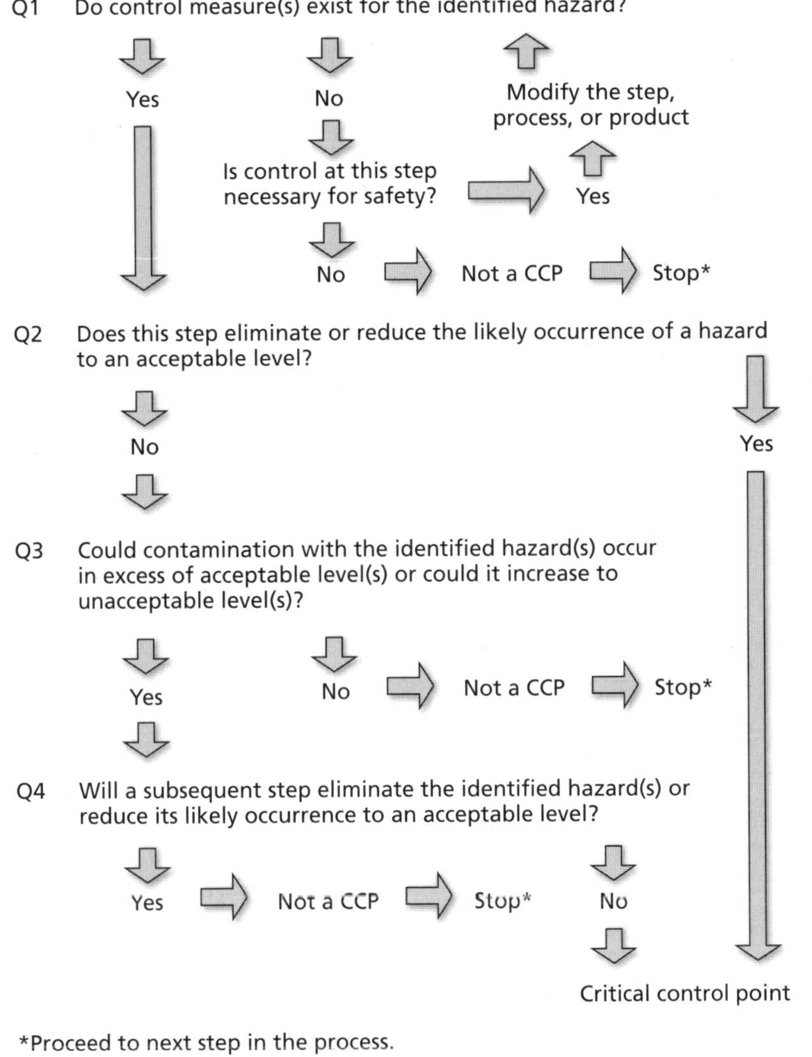

Figure 6.3 NACMCF decision tree 2.

Source: National Advisory Committee on Microbiological Criteria for Foods (NACMCF), "Hazard Analysis and Critical Control Point Principles and Application Guidelines," *Journal of Food Protection* 61 (1998): 1246.

If the answer is "no," ask whether control at this step is necessary for the safety of the product. If it is necessary, then determine how the identified hazard could be controlled before, during, or after the manufacturing process. Often a step, process, or product can be modified to add a control measure. If control at this step is not necessary for safety, the step is not a CCP. Proceed to the next step in the process.

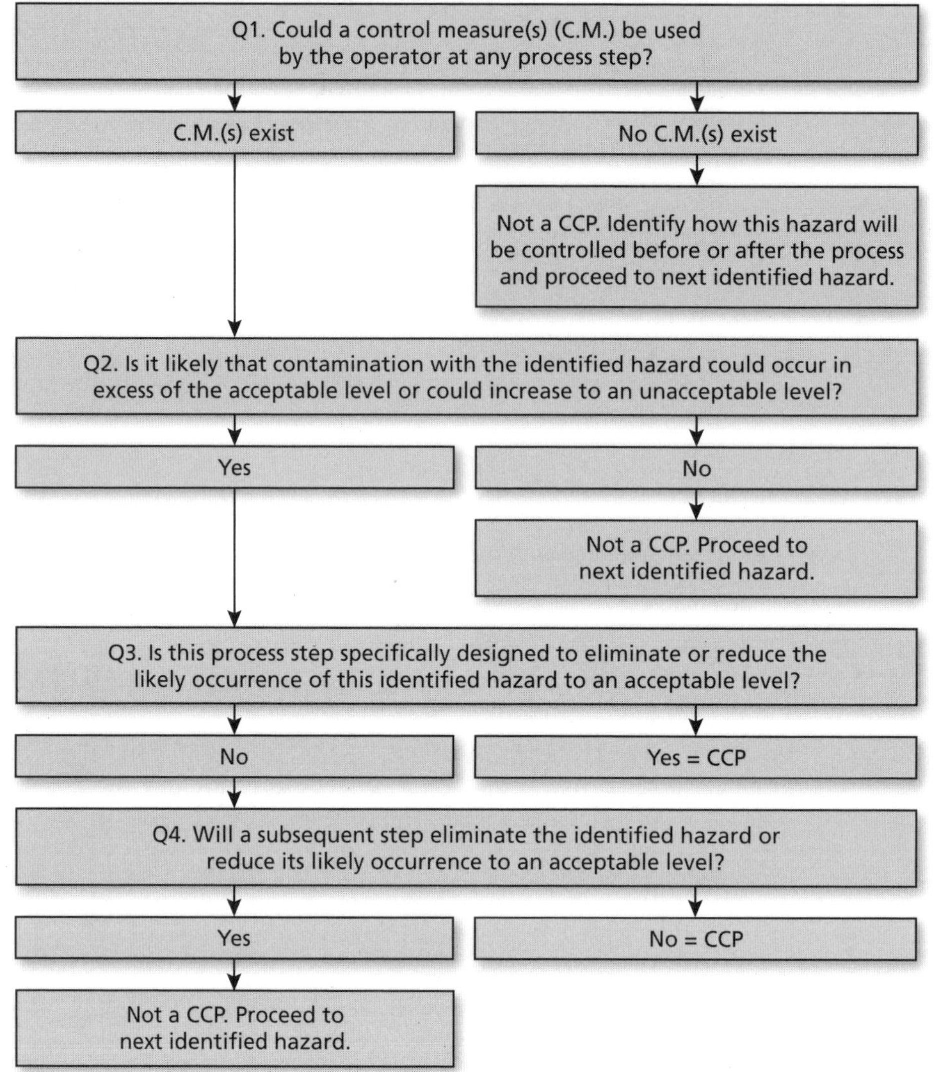

Figure 6.4 Decision tree developed by Canadian Food Inspection Agency.

Source: Canadian Food Inspection Agency, Food Safety Enhancement Program Manual, July 11, 2013, http://www.inspection.gc.ca/food/safe-food-production-systems/food-safety-enhancement-program/program-manual/eng/1345821469459/1345821716482.

Question #2: Is this process step specifically designed to eliminate or reduce the likely occurrence of the identified hazard to an acceptable level? "Specifically designed" means that the procedure or step is intended to specifically address the identified hazard. For example, a metal detector has been specifically designed to detect and reject products containing certain steel and stainless steel fragments. Other examples include pasteurization (a heat process designed to kill harmful organisms), retort (the process of canning products in a vessel—resembling an oversized pressure cooker—at a high temperature to kill pathogens), and acidification (adding acid to lower the pH of foods to make conditions unfavorable to bacterial growth).

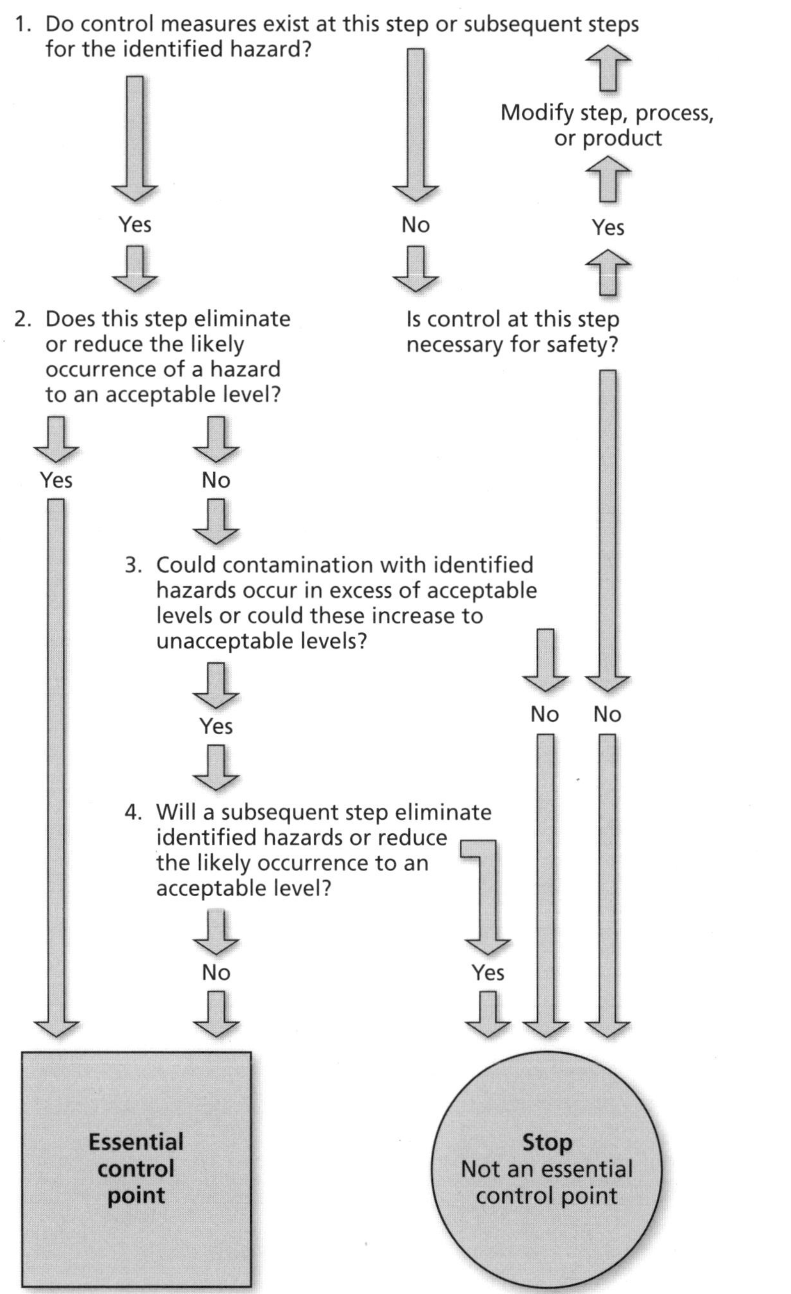

Figure 6.5 ECP decision tree table for medical devices.

Source: Medical HACCP Alliance, *Hazard Analysis and Critical Control Point* (Blacksburg, VA, 2008).

Process step	Is this hazard of sufficient likelihood of occurrence and severity to warrant its control? Yes—Proceed to next question No—Not a CCP	Is this hazard fully controlled by a PRP? Yes—Not a CCP No—Proceed to next question	Q1. Do control measures exist for the identified hazard? Yes—Proceed to next question No—Either not a CCP or need to modify step, process, or product	Q2. Is this process step specifically designed to eliminate or reduce the likely occurrence of the identified hazard to an acceptable level? Yes—CCP No—Proceed to next question	Q3. Could contamination with the identified hazard occur in excess of acceptable level(s) or increase to unacceptable level(s)? Yes—Proceed to next question No—Not a CCP	Q4. Will a subsequent step eliminate the identified hazard or reduce its likely occurrence to an acceptable level? Yes—Not a CCP. Identify subsequent step. No—CCP	CCP number

Figure 6.6 CCP determination form.

A Is this hazard of sufficient likelihood of occurrence and severity to warrant its control?

Yes

No → Not a CCP → Stop*

B Is this hazard fully controlled by a PRP?

No

Yes → Not a CCP → Stop*

Q1 Do control measures exist for the identified hazard?

Yes

No

Is control at this step necessary for safety? → Yes → Modify the step, process, or product

No → Not a CCP → Stop*

Q2 Is this process step specifically designed to eliminate or reduce the likely occurrence of the identified hazard to an acceptable level?

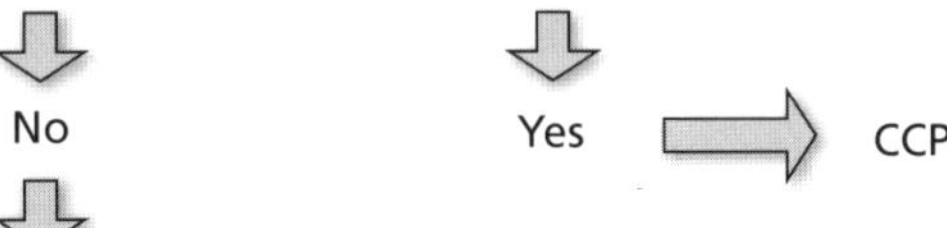

Q3 Could contamination with the identified hazard occur in excess of acceptable level(s) or increase to unacceptable level(s)?

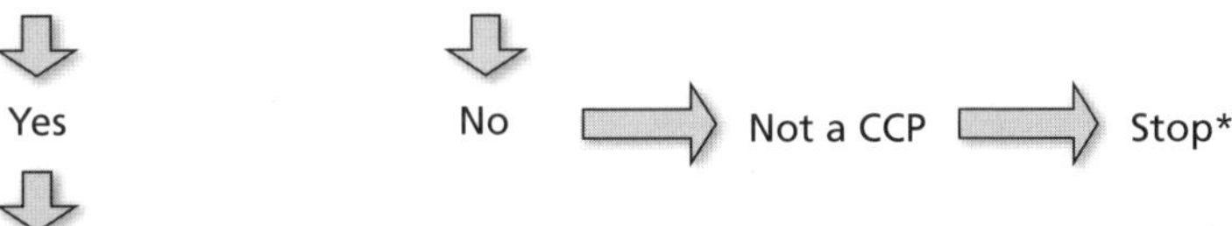

Q4 Will a subsequent step eliminate the identified hazard or reduce its likely occurrence to an acceptable level?

*Proceed to next step in the process.

Figure 6.7 A CCP decision tree.

Source: Adapted from Sara Mortimore and Carol Wallace, *HACCP: A Practical Approach,* 3rd ed. (New York: Springer, 2013).

If the process step is specifically designed to eliminate the hazard or reduce its likely occurrence to an acceptable level, the answer is "yes," and this step automatically becomes a CCP. Caution should be taken when more than one step in the process has been specifically designed to eliminate or reduce the hazard. In that case, only the final control measure is the CCP. For instance, in a process with several magnets or metal detectors in the same line, only the final metal detector is the CCP. The other control measures are control points.

If the step is not specifically designed to eliminate or reduce the likely occurrence of an identified hazard, the answer is "no." Proceed to question #3.

Question #3: Could contamination with the identified hazard occur in excess of acceptable level(s) or increase to unacceptable level(s)? This question is included in all decision trees. If the significance of the hazard has been carefully evaluated as part of a hazard analysis, the question has already been answered. Basically, this question assists the HACCP team in determining whether the hazard has a significant impact on product safety.

It is important to note that this question also asks if the hazard could increase to an unacceptable level. The growth of pathogenic microorganisms during processing, storage, or distribution could result in microbial numbers increasing to an unacceptable level even though these numbers were acceptable at an earlier point in the process.

If the answer to this question is "yes," then proceed to question #4. A "yes" response should be accompanied by a documented description of the scientific literature or data on which the decision was based, particularly if this was not done during the evaluation stage of the hazard analysis.

If your answer is "no," then the hazard is not known to have an impact on product safety, and the step is not a CCP. Proceed to the next step in the process.

Question #4: Will a subsequent step eliminate the identified hazard or reduce its likely occurrence to an acceptable level? This question assists in the identification of multiple points at which control measures may be used to control a hazard in a single process. In this case, only the final point is considered a CCP; the others are considered control points. For example, in the making of marinara sauce, the washing of tomatoes is a control point. Thermal processing is a subsequent step that will "eliminate the identified hazard." Therefore, the time and temperature parameters for the thermal processing are the CCPs. During the production of an electronic circuit for a medical device, each of the components can be tested individually prior to placing them on the circuit board. The testing of each of those components is a control point. The final electronic circuit board is functionally tested after the assembly operation. That test is considered a CCP.

If no subsequent steps are scheduled in the process to control the hazard, answer "no," and this process step becomes a CCP. If there are subsequent steps later in the process that will eliminate the identified hazard or reduce it to an acceptable level, answer "yes," and this step is not a CCP. As part of its CCP documentation, the HACCP team will need to identify the subsequent steps that control the hazard before proceeding.

Factors Leading to CCP Misidentification

A number of factors can lead to the misidentification of CCPs, including the following:

- Using a decision tree without first determining the hazard's significance and whether it is controlled by a PRP
- Missing a process step in the process flow diagram and hazard analysis
- Failing to identify all possible hazards
- Assigning the wrong level of significance to a hazard
- Inadequate development or implementation of PRPs
- Misapplication of the decision tree
- Lack of scientific evidence in support of hazard identification

DOCUMENTING CRITICAL CONTROL POINTS

CCP determination should be clearly documented. Figure 6.6 provides an example of the type of form used. CCPs can be sequentially numbered on the process flow diagram, CCP determination form, and HACCP plan for convenience. Many organizations identify CCPs numerically with a category qualifier for biological (B), physical (P), and chemical (C). For example, if the first CCP identified controls a biological hazard, it is recorded as CCP-1B. If the second CCP identified controls a chemical hazard, it is recorded as CCP-2C. If both biological and chemical hazards are controlled at the same processing step, and this is the fifth CCP, then the CCP number used is CCP-5BC. This identification protocol sequentially identifies CCPs independent from process step numbering and informs the user of the HACCP plan which types of hazards need to be controlled at a particular process step.[7]

NOTES

1. National Advisory Committee on Microbiological Criteria for Foods (NACMCF), *Hazard Analysis and Critical Control Point Principles and Application Guidelines* (Washington, DC: U.S. Food and Drug Administration, August 14, 1997).
2. Association of Food and Drug Officials (AFDO), *Medical Device HACCP Training Curriculum,* draft ed. (York, PA: AFDO, 1999).
3. NACMCF, *Hazard Analysis.*
4. Sara Mortimore and Carol Wallace, *HACCP: A Practical Approach*, 3rd ed. (New York: Springer, 2013): 113.
5. Ibid., 214.
6. Kenneth E. Stevenson and Dane T. Bernard, *HACCP—A Systematic Approach to Food Safety,* 3rd ed. (Washington, DC: The National Food Processors Institute, 1999): 82.
7. Canadian Food Inspection Agency, *Food Safety Enhancement Program Implementation Manual,* Vol. II (Ottawa: Government of Canada, 1995): 57–58.

Chapter 7

Principle #3—Establish Critical Limits

WHAT ARE CRITICAL LIMITS?

The third HACCP principle of setting critical limits follows two very important stages in the development of a HACCP plan. The critical limits are determined after a thorough hazard analysis has been conducted and the correct critical control points (CCPs) have been determined. The NACMCF defines a critical limit as "a maximum and/or minimum value to which a biological, chemical, or physical parameter must be controlled at a CCP to prevent, eliminate, or reduce to an acceptable level the occurrences of a food safety hazard."[1] The Codex Alimentarius Commission defines a critical limit simply as "a criterion which separates acceptability from unacceptability."[2] In other words, critical limits are defined processing boundaries that cannot be exceeded. If a critical limit is exceeded, then the product must be considered potentially unsafe and corrective action must be taken.

Parameters commonly used in establishing critical limits and controlling biological hazards include time, temperature, weight/size, humidity, water activity (A_w), pH, preservatives, salt level, chlorine level, and viscosity.

An example of setting critical limits occurs in the cooking of meatballs. The critical limit for microbial destruction in a meatball is a combination of time and temperature.[3] A critical limit could be set by cooking the meatballs to an internal temperature of 160°F (71°C) or greater. At this temperature, there would be a 7 $\log_{10}$ reduction in microorganisms in less than one second. Because of the very fast reduction in microorganisms, there would be no need to record the cooking time. If lower temperatures were used, then the meatballs would have to be cooked for a longer time, and both the internal temperature and length of cooking time would need to be recorded to ensure the pathogenic microorganisms were destroyed.

Another example of setting a critical limit occurs with the cooling of a cooked meat product. The maximum internal temperature of a cooked product should not remain between 80°F (27°C) and 130°F (54°C) for more than 1.5 hours, nor between 40°F (4°C) and 80°F (27°C) for more than 5 hours.[4] This ensures that recontamination of the cooked product and exponential growth of bacteria do not occur.

ESTABLISHING CRITICAL LIMITS

A scientific basis needs to be used to set critical limits. To do this, companies may utilize external sources of information to augment internal sources of knowledge. Sources of information to use when setting critical limits for biological, chemical, and physical hazards may include literature searches, government regulations, industry standards, trade association technical committees, in-house research or studies, generic food safety models, equipment manufacturers, and trained industry consultants. Once the critical limits are established for a CCP, processing limits can be developed to control the process more tightly or simply to keep it within normal boundaries. Critical limits should define unacceptable processing conditions from a safety perspective. If the critical limit is exceeded, corrective actions must be initiated. Critical limits differ from operational limits. Operational limits define normal processing conditions and should be set more tightly than critical limits.

Microbiological

Microbiological finished product testing or performance standards are rarely used as critical or operational limits simply because of the delay between the time of sampling and the receipt of testing results. An example is the USDA/FSIS performance standard of 6.5 $\log_{10}$ reduction of salmonella for cooked poultry products. This is to ensure that cooked chicken will not contain salmonella, a bacterium that causes foodborne illness.

Sampling of the product for microbiological issues can be rather ineffective, especially if the pathogen is present at a low level or is not randomly distributed throughout the production lot. A better method to ensure microbiological safety is to define the processing conditions needed to achieve specific time/temperature parameters, thus ensuring that the microbiological performance standards are met. Controlling the process allows immediate corrective actions to be made if a process's critical limits are exceeded so that risks resulting from the production of a product are minimized or averted.

Chemical

Chemical limits take a number of different forms. Chemicals can be both naturally occurring and added. Examples of naturally occurring chemical hazards include shellfish toxins, aflatoxins, and vomitoxins. Critical limits for shellfish toxins can relate to the time of year of harvesting shellfish and the harvest water location. For vomitoxins, which occur in wheat, corn, and other grains, grain from an entire region can be affected or the problem may be localized. The incidence of vomitoxins is related to weather conditions. In this case, regular testing of the supply source must take place prior to use of the grain by a flour mill or grain mill.

Other naturally occurring chemical hazards, such as lead, mercury, or even dioxins, can come from a contaminated environment. These can be controlled, for example, by a critical limit that specifies "no lead as provided by a supplier source guarantee." Potential chemical hazards such as pesticides, hormones, antibiotics, preservatives, colors, vitamins, and nitrites are most effectively controlled through good manufacturing practices (GMPs), good agricultural practices (GAPs), and prerequisite programs (PRPs).

Physical

Defining critical limits on physical hazards is straightforward. Equipment such as magnets, metal detectors, sifters, and screens remove many physical hazards. For a magnet, the critical limit could be described as "no hazardous ferrous metal," whereas for a metal detector, the critical limit can be based only on the metal detector's capability to find ferrous, nonferrous, and stainless steel material. The proper functioning of the kick-out device should be defined and monitored. All kick-outs should be carefully checked to investigate the source of the metal.

ESTABLISHING OPERATIONAL LIMITS

Once the critical limits are established, operational limits can be established. Operational limits are designed to prevent routine deviations from the critical limits. Operational limits are set tighter than the critical limits to provide a safety factor for the processor. These limits need to take into account various factors such as accuracy and precision of the measurement process, process and product variation, and limits needed to achieve quality requirements. If the potential exists for occurrence of a process deviation, operational limits provide the processor with the opportunity to adjust the process and bring it back into control prior to the production of a product that violates the critical limit.

NOTES

1. National Advisory Committee on Microbiological Criteria for Foods (NACMCF), "Hazard Analysis and Critical Control Point Principles and Application Guidelines," *Journal of Food Protection* 61 (1998): 1246.
2. Codex Alimentarius Commission, *General Principles of Food Hygiene*, CAS/RCP 1-1969 (Geneva, 2003).
3. USDA/FSIS, Appendix A: "Compliance Guidelines for Meeting Lethality Performance Standards for Certain Meat and Poultry Products," 1999, http://www.fsis.usda.gov/OPPDE/rdad/FRPubs/95-033F/95-033F_Appendix%20A.htm
4. USDA/FSIS, Appendix B: "Compliance Guidelines for Cooling Heat-Treated Meat and Poultry Products (Stabilization)," 1999, http://www.fsis.usda.gov/OPPDE/rdad/FRPubs/95-033F/95-033F_Appendix%20B.htm

Chapter 8

Principle #4—Establish Monitoring Procedures

WHAT IS MONITORING?

The activity of monitoring CCPs is essential to the success of a HACCP-based system. Appropriate monitoring procedures must be established and used to ensure that critical limits are not exceeded. To establish and effectively conduct monitoring procedures, the questions of who, what, where, when, why, and how must be answered. Such procedures typically are observations or physical measurements that can be readily conducted without imposing unrealistic time delays or extreme costs on production processes.

Within a HACCP-based system, monitoring is the act of scheduled testing and/or observation, recorded by the company, to report the findings at each CCP. Monitoring is an action that normally is carried out on the line by production personnel. Monitoring may be performed continuously by an instrument such as a temperature or pH recorder. Since monitoring consists of continuing observation, management attention and action are needed to sustain the HACCP process. Management must ensure that appropriate actions are taken when critical limits are exceeded. HACCP is not something that can be set up, turned on, and ignored.

Examples of monitoring procedures include:

- Sampling and inspecting raw materials
- Checking and documenting product temperatures
- Checking temperature and humidity in dry storage rooms
- Checking inventory control
- Checking amounts of additives used for each batch/lot
- Product sampling for bacterial analysis
- Scheduled checking of net weights
- Scheduled checking of labels used
- Periodic checking of process control specifications
- Visually inspecting product and equipment
- Checking equipment maintenance

The monitoring procedures used for each CCP must be specific and should be designed to monitor the control of each hazard identified.

Monitoring is done to collect data and to subsequently have information upon which to base decisions and take appropriate actions. It also provides an early warning that a process is either losing control or is in fact out of control. When done properly, monitoring can help to prevent or minimize loss of product when a process or handling deviation occurs. It can also help to pinpoint the cause of the problem when control is lost. Without effective monitoring and recording of data or information, there is no HACCP-based system.

COLLECTING DATA

Monitoring is a data collection activity. Thus, it is important to understand how to properly collect data. In general, there are 10 steps to follow in designing a data collection or monitoring activity:

1. *Ask the correct questions.* The questions must relate to the specific information needed. Otherwise, it is easy to collect data that are incomplete or that answer the wrong questions.
2. *Conduct appropriate data analysis.* What analysis must be conducted on raw data collected to allow a comparison with the critical limit(s)?
3. *Define where to collect.* Where are the specific locations for data collection?
4. *Ensure unbiased data collection.* If the data are biased, the data do not accurately describe the manufacturing conditions, and the data subsequently cannot be compared to the critical limits. Therefore, it will not be possible to determine if the products being manufactured are safe. It is critical to define both the accuracy and precision needed for specific monitoring processes.
5. *Understand the needs of the person collecting the data.* This may include special environment requirements, training, and experience.
6. *Design simple but effective data collection forms.* Forms used to collect and record data should be self-explanatory, must permit the recording of all appropriate data, and should be designed to reduce the opportunity for error. If food safety data are collected electronically, the data collection system must comply with appropriate regulatory requirements.
7. *Prepare instructions.* Operating or work instructions are powerful tools to help employees conduct the monitoring process in a consistent manner.
8. *Test the forms and instructions and revise as necessary.* The person who developed a form may think it is easy to use. However, a form may not be easy for the operator to use under production conditions.

9. *Train the person(s) collecting the data.* Operators need to know how, where, and when to collect the data and how, where, and when to properly record their observations.

10. *Audit the collection process and validate the results.* Audits provide a tool to ensure a HACCP system is: (1) operating as planned; and (2) operating in an effective and efficient manner.

Monitoring can be done either by observation or measurement at CCPs. In general, an observation gives a qualitative index of control; a measurement results in a quantitative index. Thus, the choice of whether the monitoring will be an observation, a measurement, or both depends on the established critical limit and available methods. Potential time delays and costs should also be considered.

Monitoring by Observation

Observation is the most basic means of data collection. While monitoring by measurement often is recommended because it gives "unbiased" numbers, the importance of observations cannot be overlooked. Of course, the observations must be compared to the CCP's critical limit(s). This requires a manual analysis by the observer and, in many cases, a subjective interpretation. Extreme care must be taken when selecting, training, and standardizing (calibrating) the individuals making the observations.

Monitoring by Measurement

Monitoring by measurement can include physical, chemical, or microbiological indices. The most common process measurements are time, temperature, and pH. However, for raw materials, chemical tests for toxins, food additives, and contaminants and microbiological tests for coliforms, *Escherichia coli,* salmonella, and other microorganisms are often used.

As expected, measurement monitoring requires some extra care. Equipment must be calibrated and work instructions for data collection must exist. Monitoring with an uncalibrated thermometer or one that does not read to the desired decimal point can do more harm than good.

RECORDING DATA

Multiple procedures can be used to record measurement data. The easiest way is with a data sheet. Data sheets should record information in a simple format. However, as in observation monitoring, the person recording and collecting the data must be instructed sufficiently to analyze data relative to the critical limit. In the example shown in Figure 8.1, the person collecting the data must know that any internal temperature below 180°F (82°C) is cause for a corrective action.

The trend in measurement monitoring is toward full automation. Microprocessing systems can activate visual and/or sound alarms when critical limits are exceeded. Automation can produce data sheets as well as control charts and check sheets. If calibrated and maintained correctly, automated systems can help to reduce the risk of human error.

Processing line 1 critical limit: Temperature not less than 180°F		
Time	**Temperature**	**Notes**
0800	181°F	
0830	181°F	
0900	180°F	
0930	180°F	
1000	179°F	1005—Increased steam flow
1030	180°F	
1100	180°F	
Operator: Joe Smiley		Date: mm/dd/yy

Figure 8.1 Temperature data sheet.

CONTINUOUS VERSUS INTERMITTENT MONITORING

If monitoring is not continuous, the question of when to monitor becomes extremely important. It is no less important for in-process monitoring than for monitoring on an individual lot basis. Intermittent or noncontinuous monitoring must reliably indicate that a hazard is under control.

Intermittent monitoring quickly leads to a discussion on statistics. If monitoring is on a per-lot basis (for example, raw material), the question becomes, "How much do I sample?" If a production line is to be intermittently monitored, an additional question to ask is, "How often do I sample?" These questions are best answered through statistical analysis. Management must decide on the amount of risk it is willing to accept by consulting literature or competent statistical authorities. These sources can help answer the "when" question so appropriate intervals can be incorporated into the development or modification of a firm's sampling plans.

DETERMINING MONITORING POINTS

Monitoring in a HACCP-based system is performed at CCPs. It must be done at a location where a CCP accurately reflects the state of a critical limit. For example, if the critical limit of a cooking CCP is an internal temperature of 180°F (82°C), then monitoring should be performed during or immediately after cooking when the maximum temperature has been reached. However, the ideal is to monitor where there is minimal interruption in the production flow.

With regard to other monitoring location options, the key to establishing "where" is to learn to ask the correct questions. Only then can effective data collection occur. The following questions should be asked (sample answers using the aforementioned internal temperature example are provided):

Q: What questions need to be answered?

A: What is the internal temperature of the cooked product at the completion of cooking (or after a certain amount of time)?

Q: What means will be used to get the answer?

A: Measuring the internal temperature at the cold spot in the sample.

Q: What data analysis must be done, and how will the results be communicated?

A: The internal temperature must be at least 180°F. Results can be communicated by an electronic alarm, by continuous monitoring by a temperature recorder, or by periodic examination of a temperature control chart.

Q: What type of data are needed?

A: Internal temperature data recorded to the nearest 0.1°F are needed.

Q: Where in the process can data be obtained?

A: Data can be obtained from the cooked product as it exits the cooker.

Some of these questions and answers are self-evident, and it might appear foolish to follow this process for every critical limit. However, the important concept is that, in deciding where to collect data, the process works backward. Before asking what data are needed, the company should determine what questions must be answered. This defines data need and subsequently suggests where the data should be collected.

QUALIFICATIONS OF THE PERSON COLLECTING THE DATA

The qualifications of the person collecting the data must be based on the "how" and "where" of monitoring. Certainly, the designated employee must have easy access to the CCP, as well as the skills and knowledge to understand not only the production process but also the purpose, importance, and process of the monitoring activity. In cases such as organoleptic determination of decomposition or chemical or microbiological analyses, the person must have a high level of training and experience. Of course, the person should also be unbiased. All things considered, the "who" should be someone in whom the company can place its trust.

Chapter 9

Principle #5—Establish Corrective Action Procedures

WHAT IS CORRECTIVE ACTION?

No matter how well the HACCP system has been designed and implemented, deviations from the processes established as part of the HACCP plan may occur. Once hazards have been identified, corresponding critical limits determined, and monitoring procedures set up, it is necessary to establish corrective action procedures. Corrective action is defined as "action to be taken when the results of monitoring at the CCP indicate a loss of control."[1]

It is a general sign that a process is out of control when the critical limits at a critical control point (CCP) are exceeded. When this occurs, corrective actions must be taken to bring the process back into control, and appropriate action must be taken on the product that was produced while the process was not in control. The HACCP plan must be designed so that deviations from the critical limits can be discovered quickly. The early detection and subsequent elimination or reduction of deviations enables corrective actions to be taken as early as possible. This minimizes the production of nonconforming product. Specific corrective action plans should be developed for each CCP since variations result from many causes. Activities that would be considered corrective actions include isolating and holding product for safety evaluation, diverting the affected product or ingredients to another line where deviation would not be considered critical, reprocessing, rejecting raw material, and destroying product.

Personnel responsible for implementing corrective actions must be properly trained; possess a thorough understanding of the HACCP plan, the process, and the product produced; and have the authority to ensure that the corrective action is properly taken. Additionally, the corrective action should be documented.

GOALS OF CORRECTIVE ACTION

Corrective action has two goals. The first is to identify, correct, and eliminate the cause of the deviation. The second is to determine the scope of the problem so nonconforming product can be properly identified and disposed of. These two elements apply when the deviation is associated with exceeding critical limits and should lead to restoring control of the process.

There are four general steps to any corrective action. They are to: (1) identify the cause of the deviation; (2) determine product disposition; (3) record the corrective action; and (4) reevaluate the HACCP plan. The corrective action documented in the HACCP plan can be considered a short-term corrective action plan. It is designed for immediate response. Once the short-term corrective action is taken, the company should devise a response that eliminates the root cause of the problem. This process is part of the reevaluation of the HACCP plan.

Identifying Causes of Deviations

Identifying the cause of a deviation usually involves some form of root cause analysis. A root cause is a fundamental deficiency that results in a nonconformance and must be corrected to prevent the reoccurrence of the same or similar nonconformance. The misidentification of the root cause could lead to an improper corrective action.

Determining Product Disposition

Determining the appropriate method for disposing of nonconforming product is important, as it is undesirable for nonconforming or unsafe product to enter commerce. The first step is evaluating the nonconforming product. This can be done in one of three ways: (1) by using evidence other than the monitoring system demonstrating that the control measures have been effective; (2) by using evidence showing that the combined effect of the control measures for that particular product complies with the performance intended; or (3) by using the results of sampling, analysis, and/or other verification activities demonstrating that the affected lot of product complies with the identified acceptable levels for the food safety hazard concerned. If the criteria are not met, then the company has two options: (1) reprocessing or further processing within or outside the organization to ensure that the food safety hazard is eliminated or reduced to acceptable levels; or (2) destruction and/or disposal of the product as waste. Product destruction should be witnessed and documented.

Disposal is a short-term correction designed to contain a problem. Long-term corrections address the underlying cause and are expected to solve a problem permanently without resulting in the occurrence of a new problem. To address the problem correctly, the manufacturer needs to identify the hazards and their causes. The company must improve the manufacturing processes, revise company procedures, and train employees to follow the new procedures. The processes used to determine the short- and long-term solutions should be included in the corrective action plan.

Recording the Corrective Action

All corrective actions taken must be documented. Corrective action records assist the company in the identification of recurring problems and may be used to determine whether the HACCP plan requires modification. The documentation should identify the product, describe the deviation, detail the action to be taken (including the method to be used for final disposition of the affected product), and state the name and job title of the person responsible for making the correction.

Any reevaluation of the HACCP plan should be noted in the minutes of the verification meetings.

All corrective action records should be filed separately. Additionally, corrective action records may be filed with the records that illustrate the deviation. This method of double filing permits fast and efficient evaluation of the deviations noted, and their subsequent corrections, by company personnel and outside auditing bodies.

One additional record, called a notice of unusual occurrence and corrective action (NUOCA), may be used for corrective actions. Some companies call a NUOCA a corrective action request (CAR). It is designed for use in those situations where an applicable predetermined corrective action is not available for a given scenario. This record is completed using the formal steps described previously and filed with the other corrective action reports. The acronym "NUOCA" often is used to mean any corrective action report, and in fact many companies title their corrective action reports as NUOCAs. This supports the position that corrective actions should be unusual occurrences.

Reevaluating the HACCP Plan

Many companies miss the final but extremely important step involved in corrective action: reevaluating the HACCP plan. This step can be used to: (1) identify gaps in the HACCP plan; (2) identify hazards that may have been overlooked initially; (3) determine whether corrective actions taken are sufficient to correct deviations; (4) establish whether critical limits are properly set; (5) determine if monitoring activities are adequate; (6) determine if new technologies are available that could reduce the likelihood of the occurrence of a hazard; and (7) determine if new hazards must be addressed in the HACCP plan.

CORRECTIVE ACTION PLANS

Corrective action plans usually are written in an "if–then" format. Corrective action plans should be as specific as possible, but it is not necessary for them to be extremely long. A description of what actions will be taken when a specific deviation occurs is sufficient. Following are several examples of corrective action statements:

> If temperature of milk at pasteurizer drops below the critical limit, then milk flow is diverted until temperature recovers. Diverted milk is repasteurized. Check the operation of the heating/cooling units to determine the reason for the temperature deviation that caused the flow diversion. Repair if necessary, reestablish control, and resume production. (Association of Food and Drug Officials [AFDO]—Seafood HACCP Alliance)
>
> If product does not reach required internal temperature for the required time, then recook or destroy the product. (AFDO—Seafood HACCP Alliance)

If the internal dimensions of components do not meet specifications, then reject the lot of components. (AFDO—Medical Devices)

If sterilization parameters are not met, then quarantine product, review records, and determine through sampling and testing if product is releasable, can be resterilized, or if it must be discarded. (AFDO—Medical Devices)

NOTE

1. Codex Alimentarius Commission, *General Principles of Food Hygiene,* CAS/RCP 1-1969 (Geneva, 2003).

Chapter 10

Principle #6—Establish Verification Procedures

OBJECTIVES OF THE VERIFICATION PROCESS

The complexity of the verification process makes it the least well-defined of the seven HACCP principles. It addresses several concepts under one principle. Many believe that the verification process is simply calibration and record review. While these are integral elements of verification, the principle embodies much more.

The verification principle is designed to assist a company in the accomplishment of three HACCP objectives. First, the verification process is used to ensure that the HACCP plan is working properly. In other words, it confirms that the written plan is implemented, and that the implemented plan is identical to the written plan. Second, verification ensures that the HACCP plan is valid. When used in this manner, it is a science-based review of the rationale behind each part of the HACCP plan, such as the hazard analysis, CCP determination and verification strategies, and the establishment of critical limits. Finally, verification ensures that the HACCP plan is relevant. Since the HACCP plan is not intended to be static, once developed and implemented, it must be reviewed periodically to ensure that it remains current and effective.

At a minimum, verification of the entire HACCP system should take place annually. This ensures that all plan elements are reviewed to validate the adequacy of the plan in the identification and control of significant hazards. In addition, the annual review assesses whether the HACCP plan is functioning as designed and ensures that it continues to accurately reflect the company's product and operational requirements.

TYPES OF VERIFICATION

Verification should occur at several points throughout the HACCP plan development and implementation, as well as on an ongoing basis. Several types of verification activities will be explored in this chapter. Selecting the appropriate approach to use in a specific instance is an integral part of an effective HACCP system.

Verification can be defined as methods, procedures, and tests used to determine if the HACCP system is in compliance with the HACCP plan. The HACCP team needs to decide which methods, procedures, and tests the company should perform to verify that the HACCP system is working effectively. The HACCP plan should specify these actions, state the frequency at which these actions will be performed, and identify the person(s) responsible for performing the actions. Verification uses methods, procedures, or tests separate from and in addition to those used in monitoring (Chapter 6) to determine whether the HACCP system is in compliance with the HACCP plan or whether the HACCP plan needs modification.

Verification reports should include information about the HACCP plan and the person(s) responsible for administering and updating the plan, the status of records associated with CCP monitoring, and direct monitoring data from the CCP gathered while observing the operation. Certification that monitoring equipment is properly calibrated and in working order and results of records review and any samples analyzed to verify that CCPs are under control should be included in the report. Training records of the individuals responsible for monitoring CCPs and of the HACCP team members should be reviewed and documented as well.

The concept of verification is changing in the food safety world. In 2008, the Codex Alimentarius Commission issued guidelines for the validation of food safety control measures. These guidelines make a distinction between validation and verification. These issues are discussed in Appendix C. It should be noted that this chapter takes the traditional approach to validation that is used in the Codex HACCP standard, whereby validation is considered to be part of verification.

Several categories of verification are discussed in the following sections.

Validation

The primary type of verification is the validation that the critical limits at the CCPs are sufficient for their intended purpose. Validation is the initial step in which the HACCP plan is tested and reviewed prior to implementation. This may require the assistance of external resources to identify the biological, chemical, or physical hazards that are intrinsic to the raw materials, ingredients, or processes. A scientific or technical review of the critical limits is necessary to verify that specifications set are adequate to control the hazards that are likely to occur, and, in some cases, that specifications comply with regulatory requirements. The CCPs, critical limits, and monitoring activities must be repeatedly challenged and statistically demonstrated to prevent or control identified hazards in the company's normal operations. Microbiological or analytical testing can be used effectively to verify that the process is in control and that acceptable product is being produced. The results of the validation process provide clear evidence that the HACCP plan elements adopted by the company are effective and sustainable.

Ongoing Assessment

The second type of verification is an ongoing review that ensures the company's HACCP plan is functioning effectively. A functioning HACCP system minimizes the need for extensive product sampling and testing since the appropriate preventive measures are built into the production controls. Companies rely on verification of their HACCP plan to ensure that the plan is being followed correctly, for review of CCP records, and to determine that appropriate risk management and product disposition decisions are made when deviations occur.

A schedule of verification activities is developed as part of the HACCP plan. This includes the procedures or methods to be utilized, frequency, and person(s) responsible for performing the activity. Examples of HACCP plan verification activities that should be considered as part of a HACCP program include:

- Reviewing the HACCP plan for completeness
- Confirming the accuracy of the flow diagram
- Reviewing CCP monitoring records
- Reviewing deviations and their resolution or corrective action, including the disposition of finished product
- Calibrating thermometers or other critical measuring equipment
- Visually inspecting operations to observe if CCPs are under control
- Analytically testing or auditing monitoring procedures
- Randomly collecting and analyzing samples of in-process or finished product
- Sampling for environmental and other concerns
- Reviewing consumer or customer complaints to determine whether they relate to the performance of the CCPs or reveal the existence of unidentified CCPs
- Reviewing personnel records to determine employees' amount of training in HACCP system responsibilities
- Reviewing written records of verification inspections that certify compliance with the HACCP plan or specify deviations from the plan and corrective actions taken
- Verifying the effectiveness of the PRPs
- Reviewing modifications to the HACCP plan[1]

Designated personnel must periodically review records to ensure that specific record-keeping requirements are being met. They are then responsible for documenting that this task was in fact performed and for recording the findings of the records review. Records should be complete. At a minimum, they should include the company's name and location, date and time of activity, initials of the operator, and identity of product and product code. The HACCP plan should stress the responsibility for accurate documentation and address the seriousness of falsification. Regulatory agencies may copy records to document deficiencies, provide information for agency review, document record falsification, or facilitate the next inspection.

Review of Monitoring Records

Continuous monitoring with measuring and recording devices such as automatic time–temperature equipment is optimal, as it provides a permanent record that can be reviewed and evaluated. Noncontinuous monitoring is used when a continuous approach is not feasible, as in the case of most visual examinations; monitoring ingredient specifications; measuring pH, water activity (A_w), and product temperatures; or attribute sampling. When dealing with noncontinuous monitoring, the frequency of review is critical to ensure that the hazard is under control. Random testing at a CCP or statistically based sampling may be utilized to verify effectiveness of the CCP, critical limit, and monitoring activities. Review of these records includes the determination that monitoring has been performed as specified, critical limits have been met, and corrective action has been taken when necessary. Review should ensure that actual values, rather than conclusions, are recorded, as well as the date and time of monitoring, initials of the operator, identity of the product, and the company's name and location. The signature of the person performing the review and the date reviewed should be recorded, with follow-up being performed on a timely basis, generally within one week.

Review of Corrective Action Records

The review of corrective action records determines whether all critical limit deviations have been properly addressed following the predetermined corrective action plan. Additionally, the date of the corrective action, initials of the operator, identity of the product, and the company's name and location are confirmed. The signature of the person performing the review and date reviewed should be recorded, with follow-up being performed on a timely basis, generally within one week.

Review of Verification Records

The review of verification records includes confirmation that calibration and product testing is being conducted consistent with the HACCP plan, that the process flow diagram is current, that audits are performed with the appropriate corrective action taken as identified, and that record review is being completed on a timely basis.

Calibration

Instrumentation that is used to measure, monitor, and control the parameters identified as critical limits must be calibrated to ensure the accuracy of the results. The calibration schedule should be identified for all classes of equipment, for instance, thermometers, thermocouples and temperature/humidity recorders, and feedback loops. The method of calibration, frequency at which calibration should occur, and standards to be utilized should be specified in the plan. Primary standards that are used to calibrate routinely used equipment, such as certified thermometers calibrated and traceable to national or international standards, should be identified and maintained in a controlled manner with limited access.

Testing and Analysis

As mentioned, testing and analysis may be performed during initial validation to support the selection of CCPs, critical limits, and monitoring activities. Once the plan is implemented, random samples should be collected and analyzed independently to confirm the effectiveness of each CCP in the system. The HACCP plan should specify what tests will be performed and the appropriate corrective action to be taken, including disposition of affected products, in the event of a failure.

Observation and Audit

Verification audits range in scope from observation of a company's operations or monitoring activities to an in-depth audit performed by internal or third-party resources. Verification audits should be performed periodically and may include unannounced audits. Such audits ensure that selected CCPs are continuously under control. Additional rationale for performing audits may include the determination that intensive coverage of a specific hazard is needed because of new information concerning the hazard and associated risk; when a company's products have been implicated as a vehicle of injury, illness, or associated health complaint; or to verify that changes have been implemented correctly after a HACCP plan has been modified.

Revalidation

The third type of verification consists of documented periodic revalidation, or reassessment, independent of audits or other verification procedures that are performed to ensure the accuracy of the HACCP plan. Revalidation is performed on a regular basis, for instance, annually, and/or whenever significant ingredient (raw material), product, preparation, or packaging changes require modification of the HACCP plan. The revalidation process includes a documented on-site review and verification of all flow diagrams and CCPs in the HACCP plan.

Revalidation is similar to validation in that it considers whether the plan is adequate in general rather than focusing on the plan's daily operations. It also is similar to validation in that it must be done by someone thoroughly trained and knowledgeable in HACCP principles. However, validation occurs when the HACCP plan is initially tested and reviewed prior to implementation.

Revalidation of the HACCP plan is required when there is an unexpected system failure or a significant product, process, or packaging change, or when new hazards are recognized. Changes in raw materials or ingredients, finished product distribution systems, or even the intended use or consumers of the finished product should initiate the revalidation process.

The significance of revalidation cannot be overstated. It is the element of verification that prevents the HACCP plan from becoming obsolete and therefore ineffective. In the event of a system failure or the recognition of a new hazard, revalidation of a HACCP system should be undertaken immediately. When considering changes in product formulation, packaging, and so on, the impact of these changes must be assessed prior to the change so any modifications to the HACCP plan can be made concurrent with the change.

EXTERNAL REVIEW

In specific regulated industries, the relevant governmental agencies have the regulatory responsibility to review and approve the company's HACCP plan to ensure that the system is functioning satisfactorily. This constitutes an external form of verification. It may include document review, direct observation or measurement, and/or product sampling and analysis. A HACCP system may be found to be inadequate if the plan, in operation, does not meet regulatory requirements, personnel are not performing required tasks, HACCP records are not being maintained, or adulterated product is being produced or shipped.[2]

Verification activities should be realistic within the scope of the company's operations and defined for each CCP in the HACCP plan. Over time, verification can assist the HACCP team in identifying and improving plan or operational weaknesses, eliminating unnecessary controls, and assuring continuous effectiveness of the system.

NOTES

1. National Advisory Committee on Microbiological Criteria for Foods (NACMCF), "Hazard Analysis and Critical Control Point Principles and Application Guidelines." *Journal of Food Protection* 61 (1998): 1246.
2. Ibid.

Chapter 11

Principle #7—Establish Record-Keeping and Documentation Procedures

IMPORTANCE OF RECORD KEEPING

The seventh principle of HACCP is the establishment and implementation of a comprehensive record-keeping system. Documentation provides factual evidence that a particular activity has been adequately performed to predetermined specifications. Under a HACCP system, the documentation generated must be in the form of a formal, written record that demonstrates that the activity has been performed in a timely manner and conducted in accordance with established procedures. Once a record has been created, a formal record-keeping system must be implemented. This system must establish procedures for the identification, storage, retrieval, maintenance, protection, and disposition of documents.

Some companies view HACCP record keeping as a nuisance since these records require time to record and maintain. As a result, their reason for developing and maintaining records often is solely to meet regulatory requirements. Other firms embrace HACCP records as the appropriate way of doing business and have incorporated them as an essential part of their food safety management system. In this mode, records are viewed as tools for simplifying business life, for streamlining operations, and for offering other long-term benefits to the company.

Internal Benefits

Companies that have embraced HACCP as a way of life, or as part of their food safety quality management system, recognize four internal benefits of having a comprehensive record-keeping system. First, these records provide a reasonable certainty that a company took responsible actions when product was manufactured (for example, that safety parameters were met). Second, they offer reassurance that appropriate corrective actions were taken to reformulate the products or redesign the process when critical limits were exceeded. Third, they demonstrate that products whose safety has been compromised are identified and contained, thereby preventing subsequent transfer of risk to the marketplace. Fourth, to any party reviewing or auditing the manufacturing system, a record-keeping system shows that the manufacturing process is under control. Records present a comprehensive picture that can be used to show that hazards have been detected, minimized, or controlled. This, in turn, suggests that these potential hazards no longer pose a significant risk to a consumer's health or general well-being.

Regulatory Compliance

Regulatory agencies, on the other hand, use HACCP records to evaluate the state of control and/or compliance of a manufacturing firm. Inspectors routinely make professional judgments and reach conclusions as to whether responsible manufacturing practices have been observed. A company's failure to take corrective actions when critical limits are exceeded raises substantial regulatory concerns about the product safety controls the company employs. A company that takes immediate corrective actions after critical limits are exceeded, including further preventive steps to ensure the problem does not occur again, is viewed more positively by the regulatory agency and regarded as responsible.

Indirect Benefits

The design and implementation of a comprehensive record-keeping system is a significant component of a successful HACCP system. In addition to the benefits listed previously, several indirect benefits result from a company's adherence to record-keeping procedures. First, records provide management with a mechanism for appraising the effectiveness of processing controls and procedures, and suggest trends toward noncompliance during production. Second, records can be used to evaluate personnel and to provide a foundation for an effective training program. Finally, records that are routinely shared with employees can be used to motivate them to sustain their manufacturing practices, particularly if the monitoring results show that all critical limits have been met.

TYPES OF RECORDS

Under a HACCP system, the records of primary importance are critical control point (CCP) monitoring records, corrective action records, and verification records. CCP monitoring records confirm that the critical limits have been achieved at each CCP. Corrective action records document that appropriate actions have been taken when critical limits have not been met. Verification records confirm that the HACCP plan is being followed and that the HACCP system is valid and effective. If food safety records are collected and maintained electronically, the records must be maintained in accordance with applicable regulatory requirements.

Commonly maintained HACCP system records include observations of measurements obtained during production (time, temperature, tolerances, and so on), corrective action records in response to critical limit deviations, and verification records (calibration records, laboratory analyses, annual HACCP plan reviews, and so on). These records are specifically required for certain regulated products, but numerous other records should be maintained and controlled under the HACCP system. Examples of these records include:

- Documents supporting the HACCP plan and hazard analysis
- Written sanitation standard operating procedures (SSOPs)
- Process flow diagrams
- List of HACCP team members and their qualifications
- Data supporting critical limits, including laboratory analyses
- Outline of prerequisite programs (PRPs) and preliminary steps
- Employee training records
- Supplier guarantees, certificates of compliance, certificates of analysis (COAs), or importer verification
- Shelf-life studies
- Consultant reports
- Records of laboratory analyses (for example, pyrogen testing, water phase salt, pH, microbial challenge studies)
- Packaging design and validation records

Monitoring Records

Monitoring records are the records that confirm that the critical limits at a CCP have been met. At a minimum, the following information must be included on all HACCP monitoring records: form title, company name and location, time and date, product identity (by type/model, lot number, and so on), actual observations or measurements, critical limits, operator's signature or initials, reviewer's signature or initials, and the date of review. Examples of monitoring records are shown in Figures 11.1–11.6.

Corrective Action Records

Corrective actions are taken in response to critical limit deviations. Information contained in corrective action records provides an evaluation of the actions taken, including a description of the deviation and any corrective action taken, in addition to a notation as to final disposition of the affected product. The name of the individual responsible for taking the corrective action should be included. A sample corrective action report is shown in Figure 11.7.

Raw Material Evaluation Sheet

Name of company
Address of company

Date: ______________________ Time of evaluation: ______________________

Product: ______________________ P.O. number: ______________________

Lot number: ______________________ Supplier: ______________________

Sample method and size: ______________________

Sample no.:	1	2	3	4	5
Weight:	______	______	______	______	______
Size:	______	______	______	______	______
CCP item:	______	______	______	______	______
Color:	______	______	______	______	______
Shape:	______	______	______	______	______

Certificate for CCP item (yes/no): ______________________

Operator/QC/receiver: ______________________

Reviewed by: ______________________ **Date:** ______________________

Figure 11.1 Raw material evaluation sheet.

Note: The items in **bold** are the HACCP record. In some instances it may be desirable to separate the CCP items from the control record for the non-safety items. The critical limit for this particular record is the certificate for the CCP item.

Supplier Certificate of Conformance

Name of company
Address of company

Date
Name of your company
Address of your company

Dear Mr. John Doe:

This certifies that in accordance with your purchasing specification, this shipment of (product) meets specifications for these hazards—*Supplier name, Lot number 12345.*

Yours truly,

I. B. Honest
QC director,
Name of supplier

Reviewed by:____________________ Date:____________________

Figure 11.2 Supplier certificate of conformance.

Note: Make sure the supplier meets all the specifications for the product you want on the certificate of analysis or supplier guarantee. A blanket statement does not always ensure that the product will meet your specifications.

Processing Log

Name of company
Address of company

Date: ______________________ Product: ______________________

Critical limits: ______________________ Line: ______________________

Operator: ____

Line number	Lot number	Time of day	Temp MIG	Temp recorder	Comments

Reviewer: ______________________ Date: ______________________

If critical limits are exceeded, notify shift supervisor, and separate and identify the batch involved.

Figure 11.3 Processing log.

Label Room Inspection Log

Name of company
Address of company

Date: ______________________________

Product: ______________________________

Critical limits: ______________________________

Line: ______________________________

Label room supervisor: ______________________________

Line number	Time of day	Labeling requirement (yes/no)	Presence of CCP item	Comments

Reviewer: ______________________________ Date: ______________________________

Figure 11.4 Label room inspection log.

Temperature Measurement

Instrument/Equipment Calibration Log

Name of company
Address of company

Instrument/equipment: __________

Location in plant: __________

Serial number: __________

Model number: __________

Schedule for calibration: __________

Date calibrated	Calibration results	Method of calibration	Performed by	Reviewed by

Figure 11.5 Equipment calibration log

Finished Product Report

Name of company
Address of company

Date: ______________________ Sample no.: ______________________

Vendor: ______________________ Analyst: ______________________

The results of the analyses of sample XXXX consisting of *xx* amount of samples identified as batches 1 to 4 are as follows:

Batch	APC/G	Coliforms/ 10g	E. coli/10g	*S. aureus*/G	Salmonella/ sample

Remarks:
The above sample was analyzed using methods found in the *FDA Bacteriological Analytic Manual*, Eighth Edition.

I. M. Wright
Laboratory Director
AAA Laboratories
Address of lab

Reviewed by: ______________________ Date: ______________________

Figure 11.6 Finished product report.

Corrective Action Report

Name of company
Address of company

Date: ______________________ Lot ID: ______________________

Description of problem:

____

____

____

____

Action taken to prevent recurrence:

____

____

____

____

Date problem solved: ______________________

Current status:

____

____

Supervisor: ______________________

Reviewer: ______________________ Date: ______________________

Figure 11.7 Corrective action report.

Verification Records

Verification records document that the HACCP system is valid and is being consistently implemented. Unlike daily CCP monitoring records, verification records typically are made periodically on a predetermined schedule. Verification records include:

- The HACCP plan and modifications to the HACCP plan (for example, changes in ingredients, formulations, processing, packaging, and distribution)
- Processor records verifying supplier compliance with guarantees or certificates
- Calibration records to verify the accuracy and calibration of all monitoring equipment
- Analytical records, microbiological challenge tests, environmental tests, periodic in-line tests, and finished product testing
- Audit records of in-house, on-site inspections
- Validation records of equipment evaluation tests

Figures 11.8–11.11 are examples of HACCP verification records. For the meat and poultry industry, the USDA/FSIS has specific format, review, and signature requirements.

DESIGNING A RECORD-KEEPING/DOCUMENTATION SYSTEM

The design of HACCP record systems is not specified by formal regulation. Many approaches have been successfully used to design monitoring records. Some companies prefer to use one form for each production step. Others incorporate multiple items on one comprehensive form. Regardless of the method used to design the HACCP record system, the company should review its current documentation and determine how to incorporate HACCP activities throughout the entire system. The HACCP forms must be consistent with the HACCP plan and should specify all steps required by HACCP including analyzing hazards; identifying CCPs; determining, controlling, and monitoring critical limits; taking corrective actions; and verifying results.

The successful design of a record-keeping system requires the involvement of a variety of personnel from production, quality control/assurance, and management. The resulting diversity of perspectives representing multiple disciplines creates a comprehensive picture of the entire HACCP system. Such a design also empowers employees and facilitates communication, training, and acceptance of the record-keeping system by all personnel.

Laboratory Report

Name of company
Address of company

Date: ______________________ Sample number: ______________________

Vendor: ______________________ Critical limit: ______________________

Examined by: ______________________

Remarks:
The above sample was analyzed for the presence of the CCP item using official AOAC recognized methods.

I. M. Wright
Laboratory Director
AAA Laboratories
Address of lab

Figure 11.8 Laboratory report.

Processing Research Laboratory and/or University Extension Unit

Address of unit

Date of letter
Name of your company
Address of your company

Dear Mr. John Doe:

Various published studies document that the parameters set for your processes are adequate. This supports our studies revealing that the process will meet the parameters you have set.

Sincerely,

I. M. Helpful
Processing Research Laboratory
Address of lab

Figure 11.9 Process validation letter.

Processing Research Laboratory and/or University Extension Unit

Address of unit

Date of letter
Name of your company
Address of your company

Dear Mr. John Doe:

On [date of evaluation], during a visit to your company, your test instruments were validated against our testing equipment. Test results from three production runs indicated that the parameters continued to operate as designed.

On this date, different sample sizes also were run in the equipment at your process time. These tests met your HACCP plan's critical limits.

Sincerely,

I. M. Helpful
Processing Research Laboratory
Address of lab

Figure 11.10 Equipment validation letter.

Employee Training Record

Name of company
Address of company

Employee: ____________________

Training course	Date of course

Figure 11.11 Employee training record.

Another objective in designing a record-keeping/documentation system is to provide for contemporaneous recording of events as they happen, with verification done independently by one or more persons unrelated in functions or units. Human errors may be reduced by requiring dual sign-offs, one from a person in production and another by someone in the quality department, as a verification step.

Modifying Existing Forms versus Creating New Forms

CCP monitoring records may be integrated into established record systems. Purchase orders, invoices, inventory control records, and so on can be modified for this purpose. An obvious advantage is that employees are already familiar with procedures used to fill out these records. This should reduce the amount of transitional training. Some companies are hesitant to modify current purchase orders, invoices, or inventory control records because these forms release a substantial amount of information to regulatory agencies that is unrelated to HACCP or other regulatory requirements.

Others elect to revamp their entire record-keeping system. This strategy requires a substantial investment in time to retrain some employees and redirect others to perform administrative work and technical training necessary to make the system work. The advantage to this approach is that it allows employees to understand the entire system rather than just a small step of the system or the process in which they work.

Balanced Approach versus Overkill Approach

A balanced approach in record keeping is important. Records should be designed to be simple and easy to follow, with all monitoring steps clearly indicated. Essential information that needs to be recorded should describe who, what, where, when, and how.

It is neither practical nor desirable to develop a record for every activity within the processing environment. The overkill method frequently results in companies generating too many records. This becomes burdensome to employees, limits control of the critical aspects of the process, and increases storage space needed for files.

Simple versus Complex Records

An example of a simple record is shown in Figure 11.12. A more complex variation of the same record is illustrated in Figure 11.13. Figure 11.12 contains all the information needed to check the CCP; Figure 11.13 includes the purchase order (P.O.) number to review for information needed to check the CCP at the receiving step if any problems occur. Both are adequate from the regulatory standpoint, but the second one requires more traceback review, and its accuracy cannot be determined until two records are reviewed.

Date	Product	Vendor	Amount/ quantity	Specifications	Meets specifications (yes/no)	Inspected by	Comments

Reviewed by:________________________ Date: ________________________

Figure 11.12 Simple CCP receiving record.

Date	Product	P.O. number	Specifications	Inspected by	Pass/fail	Comments

Reviewed by: ________________________ Date: ________________________

Figure 11.13 More complex CCP receiving record.

There are many advantages to developing a HACCP record-keeping system that is simple and concise. Well-designed, simplified records provide assurance that records generated during monitoring will be focused and timely. They also reduce the time required for management review and can reduce documentation errors. While most CCP monitoring records and some SSOP records require the entry of an actual observed value (for example, 45°F, 150 ppm, 35 seconds), many records can be simplified through the use of "yes/no" or "pass/fail" entries.

Computerized versus Manual Records

Some firms use computerized records to document their HACCP activities. Certain processing activities lend themselves to computerized formats. Using computerized records to monitor CCPs, however, can be problematic. The absence of key personnel to maintain the electronic documentation system can make it difficult. Further, computerized records require an electronic data collection system as the activity is observed. One way to alleviate this problem is to collect the data on wireless handheld computers. These handheld computers permit the collection of data directly from the processing line. All computerized data collection for HACCP systems needs to be compliant with FDA guidelines on electronic record keeping.[1] USDA/FSIS has similar requirements on electronic record keeping. Finally, all data collection systems (both manual and computerized) must be designed to prohibit the repetitive or falsified recording of values known as "dry labbing" of data in the absence of actual observations.

While these problems certainly can be addressed by operational improvements, computerization of HACCP records must be carefully designed to ensure contemporaneous documentation of events, adequate security for records maintained within the system, adequate storage capacity for records, and verification of authorized electronic signatures prior to product releases. After computerizing the HACCP activities, whether in full or in part, an active review of the process—from beginning to end—is advisable to ascertain whether adequate controls have been put in place for the automated data collection in the HACCP system.

PREVENTING DOCUMENTATION ERRORS

All quality and HACCP systems require human involvement. It is common to find errors across the entire spectrum of documents required within a processing operation. Common documentation errors include the lack of a specific monitoring record at a CCP, such as the temperature of a cooler at a storage CCP, or the safety of water under an SSOP record. These types of errors usually are a result of monitoring activities that may have little variation in actual value. As practices become routine in cases like this, very little analysis is conducted, or very little thinking goes into recording observations. As a result, errors are easier to make.

Significant errors can occur simply because the specifics of the HACCP plan are not being followed. These errors include improper monitoring frequencies for monitoring activities, untimely review of records, lack of corrective action records, and "dry labbing" of data.

Effective means for minimizing errors in documentation are the implementation of a comprehensive training program and the timely review of records. Training should encompass the specifics of monitoring (who, what, when, why, how, where) and the significance of the monitoring value. This not only ensures accurate records but also provides for timely corrective action when critical limits are exceeded. A regularly scheduled training program should be a continuing activity within each processing facility. However, this must be coupled with timely review of records and audits from quality department personnel or management.

NOTE

1. *Code of Federal Regulations,* title 21, part 11.

Part III

Implementing HACCP

Chapter 12 HACCP Plan Implementation and Maintenance

Validation = collect, evaluate, technical & scientific info to determine if the HACCP plan, if implemented, will control any hazards identified.

Intermittent monitoring — when not possible or practical to measure continuously

Chapter 12

HACCP Plan Implementation and Maintenance

SUPPORTING STRUCTURES FOR HACCP IMPLEMENTATION

The implementation and maintenance of a HACCP system requires effort and commitment comparable to that exerted for any major organizational change or continuous improvement initiative. While the drivers of the HACCP initiative may reside primarily in the company's technical community, HACCP plan implementation and maintenance is ultimately multidisciplinary in nature. It requires careful planning, deployment of resources, ongoing monitoring, and constant adjustment. Quality practitioners will recognize the opportunity for application of W. E. Deming's plan-do-check-act (PDCA) principle in HACCP plan implementation. With the publication of ISO 9001:2008 (Quality management system—Requirements), the standard and its subsequent modifications adopted the concept that the PDCA cycle is more than a continuous improvement cycle. It can be used as a model for a business cycle: to develop a food safety plan (plan), implement the food safety plan (do), monitor or verify its effectiveness (check), and take appropriate actions as a result of the checks (act). In addition, modern project management methodologies are well-suited for coordinating the various resources and tasks required for successful implementation.

Auditors of HACCP plans should seek evidence that the extensive planning and organizing that precedes successful implementation has been conducted. An appropriate HACCP plan is effectively sustained as new knowledge is brought to light, as the environment dictates, and as the organization changes.

Management Commitment

Ethical and environmental factors make product safety the highest business priority, but senior management may feel secure that existing quality systems alone can meet this objective. However, the HACCP approach to product safety can be justified as a means to effectively strengthen or supplant less comprehensive or effective product safety methodologies. As discussed in Chapter 2, upper management's understanding and commitment are essential for ensuring company-wide support for and commitment to HACCP plan implementation.

Calibration Standards - Primary Standards

The core team may first identify the need or the relative merit of implementing a HACCP plan. These reasons may include:

- HACCP is required by regulation in some sectors of the food industry, for example, meat and poultry operations, where USDA/FSIS regulations require HACCP systems. In addition, FDA regulations mandate HACCP for the seafood industry and the vegetable and fruit juice industry. With the advent of the Food Safety Modernization Act (FSMA), the FDA is generating the regulations that will ensure all food processors have food safety plans that conform to HACCP concepts. Internationally, EU regulations require that all processed foods be manufactured in plants that have implemented HACCP.
- Customers often require formal HACCP plans regardless of regulatory imperative. The HACCP "convention" has been increasingly defined, refined, and promoted by various international technical bodies and advisory committees and is gaining widespread recognition as a superior means of assuring product safety. An example is the NACMCF, which published the first HACCP criteria in 1992 and revised them in 1997. Its "seven principles" approach has developed into a common language. This and other organized "bodies of knowledge" create the need for producers to adopt the convention to demonstrate appropriate technical sophistication in their approach to product safety. The adoption of HACCP may simplify and ease communication with discriminating customers.
- Internal experts identify the need to adopt HACCP. This may be driven by a realization that existing product safety measures are inadequate or inconsistently applied. An industry or company may, for example, be faced with mounting technical evidence that its products pose a unique, previously unidentified, or elevated safety risk. Inspection-oriented systems may not adequately reduce the risks. Finally, inspection-oriented systems to assure product safety may result in a high incidence of internal failure—and the high costs associated with detecting failures after they have occurred. The "problem prevention" nature of HACCP can be demonstrated to be more cost effective than inspection- and testing-oriented schemes for controlling product safety.

HACCP entails extensive planning, commitment of resources, and new transaction disciplines (monitoring, record keeping, audit/verification procedures, and trend analysis techniques). Senior management must agree with the need for HACCP and fully endorse the initiative so sufficient time and money are allocated (to assure equity with competing demands). Upper management commitment to HACCP ensures that middle-level and operational managers place proper emphasis on the transactions necessary for implementation. Visible upper management commitment also indicates to the company that the benefits realized justify the outlay of resources.

Record Retention — Mandated by Regulatory Bodies or based upon Shelf life of product

Product Safety Policies and Objectives

An effective means of demonstrating company-wide commitment to product safety is the development of a formal product safety policy. Such a policy is developed by upper management and draws cross-functionally upon the company's technical, regulatory compliance, and legal experts to ensure accuracy and precision. Upper management has the responsibility of communicating the company's commitment to food safety to all employees. HACCP is a tactical component of a comprehensive product safety policy.

An effective HACCP-based product safety policy will include statements of objectives that:

- Designate product safety as a top business priority and make the distinction that product safety is not negotiable (unlike product quality level).
- Endorse the implementation of prerequisite programs (PRPs) to support product safety.
- Specifically support the development of appropriate HACCP plans.
- Endorse and name a cross-functional steering team of technical resources and subject matter experts to train and educate others throughout the company. The team coordinates the resources necessary to implement HACCP and the supporting PRPs. This steering team should, in turn, appoint a leader or overall HACCP coordinator.
- Establish timelines for program implementation.

The steering team responsible for the implementation of HACCP will develop the specific objectives and ultimately establish company-wide responsibilities for all aspects of the HACCP program's design, development, and implementation. A typical summary of responsibilities for implementation of a HACCP system is shown in Figure 12.1.

The specific responsibilities assigned will vary from company to company, but cross-functional involvement is essential. Many organizations recognize the value of involving the sales/marketing and business development functions in the early stages of HACCP plan development. These functions are especially useful in providing insight into the intended use of the finished product, an important factor to consider when performing an effective hazard analysis.

HACCP TEAM FORMATION AND TRAINING

Effective company-wide implementation of HACCP is a multidisciplinary task that requires an effort equal to that supporting any major organizational change or improvement initiative. Team formation and training are important aspects of HACCP implementation.

Policy and objectives development
Primary responsibility: management. Support: R&D, quality assurance, marketing/sales.

HACCP plan procedure development
Primary responsibility: quality assurance. Support: R&D, operations.

Approval of procedures
Primary responsibility: management. Support: R&D, marketing/sales.

HACCP training and education
Primary responsibility: quality assurance (identifies subject matter experts). Support: R&D, management.

HACCP plan implementation
Primary responsibility: operations (HACCP team). Support: R&D, quality assurance.

HACCP plan verification (including initial scientific validation)
Primary responsibility: quality assurance. Support: R&D, operations, management.

HACCP plan improvement and revision
Primary responsibility: R&D, quality assurance. Support: operations (HACCP team).

Verification of a fully functioning HACCP system
Second- and/or third-party verifications will need to be scheduled. Primary responsibility: quality assurance. Support: R&D, operations.

Figure 12.1 HACCP system implementation responsibilities.

Team Formation

A combination of cross-functional and natural work teams is the best formation to thoroughly implement HACCP. For larger companies, a hierarchy of at least two tiers of teams is recommended in HACCP implementation.

Steering Team

The first tier is the steering team, comprising staff-level management and technical/subject matter experts. A technical leader, or HACCP coordinator—usually selected from the company's quality, research and development, or compliance group—assumes a leadership role and assembles the core team.

Organized to initiate full-scale HACCP implementation, this team is responsible for identifying the need for implementation and for garnering visible upper management commitment to the initiative. This steering team should assist in the development and communication of an overall product safety policy with HACCP as an essential component. After initial implementation, the steering team monitors the regulatory and competitive environment for new knowledge related to product safety and incorporates that knowledge into control plans.

The steering team also is responsible for organizing others for the actual implementation. These responsibilities may include forming or appointing operations teams (product teams, plant teams, or site teams) to be trained and to ultimately take responsibility for on-site HACCP implementation, and identifying internal and external experts to assist in the development of HACCP-related bodies of knowledge and to deliver training.

Operations Teams

The second tier may consist of the site-specific operations teams (product teams, plant teams) assembled, trained, and empowered to implement HACCP at the establishment level. The steering, or core, team and senior management should work together to select an operations team that includes personnel from cross-functional and natural work team areas. The selection of these individuals is among the most important considerations in implementing HACCP. Cross-functional representatives will include site quality assurance, engineering/maintenance, and production management personnel. As with the company-wide core team, the cross-functional perspective is important in assuring technical integrity of the initiative at the establishment level. A site coordinator, often a quality assurance professional, should be appointed to lead the site team.

The "natural work team" element refers to the inclusion of appropriate "process owners" on the operations team. These process owners often are line supervisors and/or operators, and their inclusion helps ensure that operational details are addressed in the development of plans. Organizations that have developed ISO 9001–based quality systems and subscribers to total quality management (TQM) recognize the value of including trained operating-level personnel in the development of work instructions. Likewise, the inclusion of these employees in planning for HACCP system implementation benefits the entire organization. Operating team membership criteria may differ by organization or industry segment, but ultimately team makeup must ensure that members are able to understand and apply the significant, scientifically based body of knowledge upon which successful HACCP implementation builds.

Team Training

Upon auditing a HACCP plan or company-wide implementation, an auditor should seek evidence that training plans are complete, and that training objectives are indeed realized and documented. Training must be well-conceived and appropriate to the company's HACCP implementation plans. A company implementing HACCP will need to demonstrate that appropriate competencies have been assured in the design and execution of training.

Extensive team training is essential to support HACCP implementation. When planning for training, the organization should first seek to understand specific external requirements. Certain regulated industry segments specify training requirements. For example, both seafood and meat and poultry HACCP regulations require training in HACCP for a designated individual in a food company. Logically, this person would be the HACCP leader or coordinator for the company.[1] Increasingly, customer requirements include specific training requirements. Obviously, these and any other requirements must be met. Records must be kept of any training provided to assure customers and provide evidence to auditors that appropriate training has indeed occurred.

Significant consideration must be given to the depth of training that is appropriate to the organization's HACCP implementation. While external requirements (regulations, customers) often require the training of coordinators, or single-initiative leaders, effective implementations are assured through appropriate training of a "critical mass" of individuals responsible for overall implementation, execution, and maintenance of the HACCP plan. Line supervisors, quality control staff, line operators, and individuals involved in the purchasing and distribution of goods should be trained to the appropriate degree since these individuals are essential to effective site teams. Each organization will address the question of training depth differently and must consider the costs and benefits of extensive training. Training should be considered a prudent preventive quality cost.

Training must be formal and may require the use of external expertise, at least in the development of initiative leaders. Many companies mistakenly interpret the considerable body of knowledge related to HACCP and rely solely on internal agents to draft HACCP plans without any outside training and influence. To successfully implement HACCP, an organization should carefully consider all available resources when designing an appropriate training intervention to support HACCP implementation. Industry trade groups and associations (composed of both primary producers and suppliers), regulators, academia, and private sector consultants—used individually or collectively—provide potential resources to draw upon for training. The use of these external experts ensures that potential microbiological, physical, and chemical hazards of public health significance are fully considered in the development of the company's HACCP plan. External experts also provide perspectives on the dynamic regulatory and consumer environment. However, an external agent's efforts must be integrated with company-specific activities to assure local buy-in and relevance.

Many organizations define the qualifications required to become an external HACCP consultant prior to awarding a contract for training. Desired qualifications include a demonstrated understanding of and subscription to the seven principles of HACCP as outlined by NACMCF (or Codex Alimentarius Commission), a demonstrated ability to prepare HACCP plans, industry experience and product category knowledge specific to the industry under consideration, and prerequisite knowledge of good manufacturing practices (GMPs), product safety, and sanitation standard operating procedures (SSOPs). Additionally, actual business references and access to a contemporary client base may be required, as is appropriate professional liability insurance. Finally, consultants must demonstrate an understanding of recognized auditing techniques.

Ultimately, the selection of external experts/consultants to assist in the training of the HACCP implementation team is the responsibility of the organization's overall HACCP coordinator. Specific details on training delivery and expectations should be developed. Additionally, the need and policies for periodic training refresher courses—or retraining—should be identified.

Training interventions must be carefully designed to ensure participants receive the most for resources expended. In-house training delivered by a qualified training consultant (as outlined previously) often is an excellent approach. Training should include initiative leaders as well as line-level participants. The focus of the training should be on company-specific applications (products, unit operations, and so on). A two- to three-day in-house course often is used by organizations implementing HACCP. The expenses and logistics associated with in-house training must be weighed against the relative merit and convenience of attending off-site professional training courses.

Off-site training often is appropriate, provided participants are exposed to material specific to their company's area of operation. Off-site training, such as public course offerings, is often cost-effective and allows participants the opportunity for the interchange and infusion of ideas that comes with exposure to other organizations. When off-site training is chosen, the training organization should be selected according to the criteria outlined previously. Off-site seminars of two to three days in length should provide participants with the opportunity to select exercises and applications to the industry segment under consideration. Ideally, off-site or public courses will provide participants with useful reference materials and "after session" opportunities to clarify any questions or considerations that may follow training.

Off-site short courses can be used to train internal trainers, who should in turn bring acquired knowledge in-house for delivery. Off-site courses may be regarded as "train the trainer" opportunities, provided participants in off-site courses are competent to bring acquired knowledge in-house and, in turn, train participants accordingly. HACCP training plans ultimately must ensure that a critical mass of individuals is appropriately prepared to implement and maintain the HACCP initiative. Finally, regardless of the training method used, training and reference materials should be assembled and made available for future reference and consultation as necessary.

In some cases, expert training in specialized subject matter may be needed. For example, advanced, specialized training in scientific validation and verification of HACCP plans may be necessary in certain circumstances. Trade and regulatory affiliated organizations increasingly offer such courses. This type of training is advised in situations where called for by close regulation or customer requirements and should be targeted for HACCP initiative leaders.

Many companies embrace the use of commercially available software products to aid in the implementation of HACCP plans. While software may be useful in assuring adherence to certain conventions in flowcharting, analyzing hazards, and so on, software is no substitute for training in the underlying principles and concepts essential to the effective implementation of HACCP. Software programs that adhere to the seven underlying HACCP principles are a useful aid in deploying HACCP, but they cannot replace extensive training of personnel.

Many small companies face practical problems associated with the costs of training to support HACCP implementation. State or local initiatives that provide training and development funds for organizations in need may be consulted for assistance, especially in cases where the implementation of HACCP is essential to company competitiveness and employee retention. Small companies should consider identifying such sources of funding and applying to secure them.

PILOT PROJECTS/OPERATIONAL QUALIFICATION OF HACCP PLANS

Failure to link the training of operations/site teams with a relevant and immediate application is a common mistake. Training should be immediately followed by the opportunity to apply principles learned at the establishment level. Ideally, a manageable pilot project—also called a demonstration project or trial—will be identified and integrated into the training that precedes HACCP implementation. The execution of a pilot project integrates training and preparation into the actual HACCP implementation. Training should conclude with the development of detailed work instructions that are relevant to actual on-site application. This enables training participants to "take home" relevant operational plans that apply to actual on-site applications and immediately design a HACCP plan, at least on a trial basis.

Pilot projects provide an organization with the opportunity to apply HACCP plans on a limited basis while studying all aspects of implementation. Results must be carefully examined so that necessary adjustments can be made to ensure success when HACCP is deployed companywide. Implementation plans should list specific projects (as identified by the guiding coalition and site teams) along with timetables for their implementation. Here, trained site teams assume primary responsibility for the implementation. In multi-site organizations, at least one pilot project should be conducted per site.

Pilot project selection may be limited to a specific unit operation or a limited family of products or processes at a specific operating site. A pilot project should be regarded as a trial to ensure that the organization can execute all aspects of implementation. A pilot plan is used to confirm the following: procedures have been learned and can be applied, critical control points (CCPs) are scientifically validated as planned and can be monitored, corrective action plans can be executed, all forms and records are employed and understood, and deviations are appropriately recognized and reasons for them understood. While a viable trial or demonstration project may be limited to a specific aspect of production, it needs to be designed so as to demonstrate the organization's ability to successfully execute the seven principles in extended areas.

During the pilot project, special attention must be paid to the scientific validation of HACCP plans (this validation will be repeated as HACCP is extended companywide). Specifically, validation consists of making sure each CCP is identified and the critical limits associated with them are based on sound technical information. Monitoring of CCPs should ensure that identified hazards have been prevented, eliminated, or reduced to an acceptable level. At this time, the company must prove that scientific validation is based on supporting scientific information from appropriate authoritative sources. Such sources will differ by industry segment; specific regulation or convention may prescribe scientific validation. Evidence of scientific validation should be recorded and made part of the permanent implementation plan.

Some companies will elect to keep external experts/consultants on hand throughout the entire pilot phase. If this is not possible, organizations are encouraged to maintain access to external experts for assistance in evaluating results and making adjustments to the plan.

Pilot projects and their examination are essential to successful HACCP plan implementation. They provide the opportunity for all involved to learn and apply every aspect of the system, and to make adjustments as necessary, prior to company-wide deployment of HACCP. Many operational nuances are identified through pilot projects. Many companies widely publicize or otherwise call attention to HACCP pilot projects. For example, to call attention to the impending program and changes in operational methodology, HACCP demonstration teams may identify the location of CCPs on the production plant floor by posting HACCP-related instructions at or near the location of each CCP. The posting identifies the CCP by location and number and includes a description of the critical limits and SOPs for monitoring, response, and corrective action.

Trial length needs to be carefully considered and will differ by organization, depending on the relative complexity of the operation. Organizations often conduct HACCP pilot projects/trials as they do new product development efforts: a relatively short initial trial (with evaluation, critique, and adjustment) followed by "scaled-up" or longer experimental runs of the plan. A key to maximizing learning from pilot projects is running them for a significant length of time, under representative conditions, so as to allow the usual sources of variation in materials, machinery, and manpower to become factors.

HACCP pilot project/trial results should be summarily evaluated after a predetermined period of time or after a certain defined number of production repetitions. Site team members should verify that all activities planned were in fact executed. The site HACCP coordinator should assemble team members and operators for a rigorous critique of all aspects of the pilot project/trial. If third-party/external expert or corporate team oversight of the trial has taken place, that input will need to be assembled as well.

The formal evaluation of the HACCP pilot projects/trials should:

- Be conducted in a manner that is rigorous and systematic. It is advised that specifically designed checklists or schedules, rather than participants' notes and observations, be used to sequentially evaluate each progressive step (the seven principles) in designing and implementing HACCP. NACMCF and other sources cited provide reference material to help design the checklists.

- Result in a judgment that trials determined that the plan is either: (1) sound, and the team may move forward with company-wide deployment without making any or only minor adjustments to the plan; or (2) in need of significant improvements (procedures, retraining, and so on) to the degree that a retrial may be required.

- (Where improvements are required) provide that all indicated actions are captured, with assignment of responsibilities and completion dates for distribution to appropriate parties. If repeat trials are necessary, the evaluation should state any expected changes in timing for company-wide deployment of HACCP.

Since communication and organization-wide learning are critical, provisions should be made for a thorough review of HACCP pilot projects, at least among individuals responsible for further deployment of the HACCP initiative. Some recommend that a one-hour presentation be made to all company personnel to acquaint them with the new HACCP system and their role in producing safe products for consumers.[2]

COMPANY-WIDE HACCP DEPLOYMENT

The same cycle of preparation, execution, review, and adjustment used in pilot projects applies to company-wide deployment of HACCP. Site teams remain responsible for actual implementation of HACCP plans, but leadership and technical resources are significantly diluted as the initiative is expanded among, and more comprehensively within, operational sites. For this reason, all significant adjustments, retraining, and so on should be identified at the pilot project level.

One problem faced by implementation teams is maintaining appropriate uniformity in the content, execution, and appearance of HACCP plans when the system is deployed throughout the company. Some uniformity and standardization is assured through appropriate team training. Undesirable differences can be reduced or eliminated during analysis of pilot projects. Successful multi-unit organizations often organize regular, formal networks of HACCP coordinators that meet regularly or exchange information. Uniformity in appearance of the program (often important from a customer's point of view) can be achieved through the use of centrally administered, version-controlled documents and forms. Paper or electronic forms may be adapted to a company's specific needs. Various appendices to sources cited throughout this book also provide excellent guidelines with respect to document and record content.

HACCP PLAN MAINTENANCE

NACMCF and others similarly define verification as those activities, other than monitoring, that determine the validity of the HACCP plan and that the system is operating according to plan. An important aspect of verification is the initial validation of the HACCP plan to determine that the plan is scientifically and technically sound.[3]

In excluding monitoring from the definition and designating initial validation as an aspect of verification, the 1997 NACMCF *Guidelines for Application of HACCP Principles* helps eliminate a common source of confusion among the terms. Verification is a more comprehensive set of activities and is usually conducted by someone other than operators of the HACCP plan.

In implementing a HACCP plan, leaders and teams must provide for a comprehensive set of activities that go beyond routine monitoring and initial validation (Appendix C provides a detailed description of validation principles). Verification procedures are detailed comprehensively in Chapter 8 and are not difficult to understand. Yet organizations often fail to provide for appropriate, periodic, systematic execution of verification procedures. These should be conducted by qualified individuals, then analyzed and interpreted to determine (supported by evidence) whether the plan is operating as intended. A good system of HACCP plan verification provides precisely that, and it does so formally and with scheduled regularity.

Auditors will recognize that overall HACCP plan verification has similarities to planning, conducting, and reporting a quality systems audit of the overall quality system but with a narrower focus on systems directly supporting product safety. Since this is an audit, extensive checklists are necessary and useful to compare the plan with what is actually occurring. Numerous guidelines, such as the 1997 NACMCF *Guidelines*, and proposed schedules are available when developing verification plans. Organizations are cautioned, however, to regard this verification plan as somewhat simplified and perhaps not comprehensive enough for certain environments.

In arranging a verification plan, organizations must review regulatory requirements that may prescribe specific verification activities in a certain industry or segment. The FDA and USDA/FSIS, for example, detail certain aspects of verification requirements in those regulations that apply to their respective industry segments. USDA/FSIS issues an excellent HACCP Plan Review Checklist that is readily adaptable to a wide variety of industries.

In industries where the HACCP convention is being adopted, discriminating customers increasingly are elevating the level of scrutiny around HACCP plans. As most recognize the critical nature of verification, many now extend verification activities to suppliers. This also needs to be considered when designing verification procedures.

A very general outline of the steps taken in planning and conducting a verification audit follows:

- Schedule formal verification activities and allocate resources.
- Examine all HACCP documentation (SOPs, work instructions) and sample all pertinent records, including exceptions and deviation reports. This may be done prior to the audit through a formal request for information.
- Review initial validation and verify revalidation for any significant process change, ingredient change, system failure, and so on.
- Plan a site visit and prepare a detailed checklist. Focus on the actual observation of work (is CCP monitoring understood, done consistent with plan, and so on).
- Verify PRPs that support HACCP. This is especially critical when a significant hazard is identified during hazard analysis and control reverts to a specific PRP. Instrument or testing equipment calibration procedures that directly enable CCP monitoring should be examined as well.
- As the verification cycle progresses, review previous verification reports and determine that corrective actions were implemented and effective, and plan adjustments were made.
- Review all written directions issued by qualified internal experts and endorsed by the organization's HACCP steering team that call for the incorporation of new knowledge into the HACCP system. Verify that actions have been taken.
- Prepare a written report. Orally review the report upon leaving the site and distribute corrective action requests (CARs).
- Follow up on corrective action as indicated.

It is recommended that formal verification activities take place at regular intervals. In general, independent records reviews should be conducted at least monthly, especially during the first year of HACCP implementation. Comprehensive verification audits (including site visits and work observations) should be scheduled and results documented at least yearly. Most importantly, verification should be conducted subsequent to any major change in process, ingredient, CCP selection, or critical limit designation. Verification should include revalidation of CCPs as necessary. These guidelines are recommendations only; any schedule can be set—adherence to it is the primary goal.

Formal verification entails a comprehensive audit and is not to be confused with day-to-day monitoring activities. Parties performing formal verification procedures should be independent of the operating site reporting structure. For example, many companies assign internal verification duties to independent staff groups, such as quality assurance, to ensure independence, adequate technical orientation, and auditing acumen.

Second-party (or customer) verification is required by some companies and allows HACCP plans to be calibrated to customer expectations. Third-party (contractors independent of the company, customer, or supply chain) verification is advantageous in ensuring adequate independence from the site being audited and can provide infusion of new knowledge and expertise. External second- or third-party auditors must be adequately qualified. In addition to general auditing skills and subject matter expertise, auditors should possess qualifications similar to those outlined in the section of this chapter pertaining to selection of trainers.

EXTENDING THE HACCP SYSTEM TO THE SUPPLY CHAIN

A comprehensive HACCP system optimally extends to the company's sources of supply and into the system of product distribution. In the food industry, for example, it has become common to refer to HACCP as a program of food safety assurance that extends from "field/farm to table." Increasingly, as part of HACCP implementation, organizations have realized that the HACCP/prerequisite convention must be extended to ensure cooperation with suppliers and distributors of goods. "Each specific process (growing/harvesting, distribution, product processing, final preparation) must have its own HACCP program to ensure a final product that meets the safety and regulatory standards of today."[4] Other types of operations, both upstream (suppliers) and downstream (distributors), may pose opportunities for the introduction of potential hazards of public health significance. Operations related to temperature control of microbiologically sensitive goods or even simple storage conditions are examples of such factors.

Again, in applying the underlying premise of HACCP, hazards manifested as a result of conditions at the manufacturers rarely will be reliably identified through examination of goods or products at the point of customer (intermediate or final) receipt. Where appropriate, control needs to be exercised at the critical steps in the process. Any company whose finished product integrity depends on events in the supply chain beyond its immediate control should consider requiring HACCP plans of key suppliers and distributors. Doing so brings about the practical and problematic challenge of managing these plans. Seldom will a company be intimately involved with designing and implementing HACCP in a supplier's location. As a result, the company will be essentially limited to conducting verification audits to judge the efficacy of the plans.

As the HACCP "convention" based on the NACMCF (or Codex Alimentarius Commission) seven principles gains wider acceptance, this common language and consistency in approach will aid in providing a basis for interorganizational coordination of product safety activities. Organizations should designate a technically competent individual to monitor the supply chain and apply HACCP principles. Supply partnerships, manufacturing/distribution alliances, and the like have blurred modern organizational distinctions considerably. ISO 9001–oriented organizations accumulate considerable documentation related to purchased goods and services, including requirements for approved testing methods, certificates of analysis (COAs), and material qualification protocols. Typically, statements requiring the manufacture of materials and goods under a HACCP plan are contained in this type of supplier-related documentation. As part of the due diligence and prequalification of suppliers, HACCP plans and hazard analysis documents should be obtained and reviewed. Ideally, firsthand observations of supplier and distributor activities and site inspections of suppliers should include examination of HACCP plans, and especially observation of work related to the monitoring of CCPs. HACCP-related audits should be performed against standards outlined by NACMCF. It is becoming more common for customers to require that their suppliers be compliant to an audit scheme that has been benchmarked by the Global Food Safety Initiative (GFSI). All the GFSI-recognized audit schemes are based on the HACCP principles.

NOTES

1. Donald A. Corlett Jr., *HACCP User's Manual* (Gaithersburg, MD: Aspen, 1998): 23.
2. Ibid., 125.
3. NACMCF, *Hazard Analysis*, 19.
4. William L. Bennet and Leonard L. Steed, "An Integrated Approach to Food Safety," *Quality Progress* (February 1999): 40.

Part IV

Auditing the Food Safety System

Audit plan (developed by lead Auditor

" evidence

" Criterion = evidence is compared to Requirements
↳ Provides required sources for the audit

" finding.

" Scope = type of Audit

Only Client can authorize audit or request
↓
defines purpose & scope of Audit & Audit objective.

Requirements = planned arrangements against which the performance of an activity is measured.

Chapter 13

Auditing Fundamentals

Auditing is one of the essential parts of the HACCP verification process, and audit results provide input to the corrective action process and management review system. HACCP audits are a series of planned activities that provide a systematic and independent examination of the product/process safety management system. Audit findings determine if the system is operating in an effective and efficient manner and assess whether the organization is capable of meeting specific food safety requirements and goals.

HACCP AUDITING AS A PRODUCT SAFETY AUDITING SYSTEM

Part of this process involves evaluating the risk assessment process the company uses in hazard identification and hazard analysis and validation studies. In the food industry, risk assessment involves assessing biological, chemical, and physical hazards.

Risk assessment is not a onetime evaluation that is conducted during the implementation phase of a HACCP program. It is a continuous process as new biological, chemical, and physical hazards are identified.

For example, prior to 1982, E. coli was classified as a benign microorganism that could be used as an indicator organism to measure the efficiency of plant hygiene processes. In 1982, microbiologists identified a serotype, *E. coli O157:H7.* This serotype causes a hemolytic uremic syndrome in humans. In 1993, this microorganism was responsible for a foodborne disease outbreak that was traced to contaminated hamburger served in a restaurant chain. During this outbreak, more than 500 individual cases of infection with *E. coli O157:H7* were confirmed, and four people died. In 1997, the same microorganism was identified as the cause of a foodborne disease outbreak in fresh-pressed apple juice. Prior to that incident, foods that had a pH of less than 4.6 were thought to have a low risk of food safety problems. As a result of these and other outbreaks, numerous interventions were applied in the food chain to reduce the occurrence of this food safety issue.[1–3]

The audit scope defines the type of audit. Figure 13.1 lists a number of types of audits. The HACCP audit evaluates a food safety management system. This includes evaluation of both the HACCP plan and the prerequisite program (PRP) plan. In addition, the HACCP audit is more than just a records review. It also entails a verification that the actual food safety processes that are established are implemented and maintained in the plant.

Compliance	Financial	Product
Cyber or electronic	Follow-up	Retail product
Desk	Pest control	Sanitation
Facility	Process	System

Figure 13.1 Examples of types of management system audits as defined by scope.

Most HACCP audits include a number of elements from several types of audits. For example, a USDA/FSIS food safety office audit of a poultry processing plant is a third-party audit. This audit can be classified as a system audit and a compliance audit. In addition, a regulatory audit, such as one conducted by USDA/FSIS, can include elements of desk, facility, sanitation, or process audits.

The type of audit also depends on the client who authorizes the audit. HACCP audits can be either first-, second-, or third-party audits. The authority to perform a first-party audit originates from the target organization's management. The authority to perform second- and third-party audits comes from outside the organization that is audited.

The first-party audit should be the most comprehensive audit. This can be achieved because the organization has detailed knowledge of its processes and where problems are occurring. One of the purposes of the first-party audit is to identify deficiencies before a second- or third-party audit, and then implement corrective actions and prevent the reoccurrence of the problems. Another purpose is to determine whether the organization took appropriate action if there was a discrepancy in the documented HACCP process.

Second-party audits usually originate from contractual requirements between a supplier and a customer. They are designed to ensure that the supplier is capable of meeting the customer's requirements. The scope of these audits usually includes both the HACCP plan and the PRPs.

Third-party audits are external audits. They can be conducted by an auditing company, regulatory agency, certification body, or registrar to determine conformance to a specified standard. A third-party audit may or may not lead to certification and registration of the food safety management system. When regulatory agencies conduct audits, they are auditing to ensure compliance with laws and regulations. Noncompliance with regulations can result in a number of actions including withdrawal of inspection, product holds or recalls, and possibly legal action.

AUDIT FORMAT

The HACCP audit is a systematic, independent, and documented process for obtaining and evaluating objective evidence to determine the extent to which food safety criteria are fulfilled. The HACCP audit includes procedures to establish the effectiveness of PRPs and the HACCP plan.

The HACCP audit should include an audit of the process used to generate or modify the HACCP plan and its PRPs. This includes auditing the hazard analysis and any validation studies that were used to determine critical control points

(CCPs), control points, and critical limits. Appendix C provides a description of validation. HACCP audits should be conducted using the following steps:

- Preparing
  - Selecting the audit team
  - Developing the audit plan
  - Conducting a desk audit
- Conducting
  - Opening meeting
  - Collecting data
  - Analyzing data
  - Closing meeting
- Reporting
- Closing out

AUDIT PREPARATION

A successful HACCP audit starts with good planning. All individuals involved in the audit have specific roles and responsibilities.

The process starts with the client requesting an audit. As part of the request, the client must define the purpose and scope of the audit. The management group responsible for the audit appoints the lead auditor. The lead auditor is responsible for the effective execution of the audit.

During the preparation phase, the lead auditor communicates with the auditee to negotiate the dates of the audit. In addition, the lead auditor is responsible for selecting the audit team and arranging logistics if the team must travel. The lead auditor determines audit requirements, develops the audit plan, communicates the plan to the auditee, develops audit documents, and requests and evaluates appropriate documents from the auditee. In addition, the lead auditor prepares the final report and ensures that the audit findings are resolved.

The lead auditor resolves any problems that occur during the preparation of the audit. For example, if an auditee raises an objection over the composition of the audit team, it is the responsibility of the lead auditor to determine whether the objection warrants changing the makeup of the team or whether the original team is the correct team for the audit.

Audit teams with two or more members must have a lead auditor. The lead auditor is responsible for selecting the audit team and assigning the team members to specific audit functions. The lead auditor also may have the responsibility of mentoring auditors.

The audit team is selected based on the nature and complexity of the audit and the team members' individual technical capabilities. Each member of the audit team should be independent of the function being audited. In addition, each team member is responsible for developing the appropriate audit plans to cover his or her assigned duties.

The lead auditor can appoint technical specialists if the team needs subject matter experts. Subject matter experts must be briefed on roles and responsibilities before the audit. Audits must be conducted with technically competent personnel. Technical expertise is especially important with regard to food safety audits due to the risk of adverse health effects, which may include death. It is the lead auditor's responsibility to confirm that the audit team has sufficient food safety knowledge and process expertise to audit the specific system. The lead auditor must add subject matter experts as necessary.

Audit time can vary depending on the scope of the audit and the complexity of the organization. The lead auditor needs to understand the auditee's schedule, time, and resources for the audit. Factors that need to be considered in determining the audit time and length include:

- Number and length of interviews
- Amount of time needed to observe operations
- Number and length of meetings
- Time to review reports and analyze data
- Meals and breaks
- Travel time between sites
- Shifts

The audit plan is the key to a successful management system audit because it ensures a systematic audit. Table 13.1 defines the parts of the audit plan.

Table 13.1 Parts of the audit plan.

Part	Description
Auditee	Identifies the organization being audited.
Purpose	Defines the reason for the audit and its objectives.
Scope	Describes the operational boundaries of the audit.
Requirements	The standards used to assess the effectiveness of the food safety management system. The statement can list both the internal and external requirements.
Applicable documents	A list of the records and other data resources that will be used to judge the food safety management system. This list can include HACCP plans, corporate policies, regulatory requirements, and international standards.
Overall schedule	Lists the time and length of the following: meetings, facility tour, interviews, meals, and breaks.
Audit team members	Individuals who will perform the audit or technical specialists, including subject matter experts.
Approvals	Signatures from appropriate representatives of the client and the auditee.

The audit schedule should be as detailed as possible. For example, it should list the times when each department is being audited. This allows the employees of the auditee to adjust their schedules to allow appropriate time for the auditor. In addition, if during the conduct of the audit the audit schedule changes, the auditor should notify the auditee so appropriate changes can be made in everyone's work schedules. These changes normally are communicated through the auditee's escort.

Before issuing the audit plan, the lead auditor should contact the audit team and verify a proposed timetable for the audit. Once the logistics of the audit are completed, the lead auditor should finalize the audit plan with a letter, memo, or email.

The lead auditor must decide on an audit strategy. Typical strategies include trace forward, trace backward, discovery, and element.

The trace forward method follows the chronological progress of the product as it flows from inputs to shipping. This method may not be appropriate in some food processing plants. For example, the incoming product could be raw, ready to eat, or even live animals. These inputs may be contaminated with pathogens, resulting in the potential for cross-contamination if the output is a product that has received an intervention designed to eliminate the pathogens. If a trace forward strategy is used, it is possible that the audit team may physically carry pathogens from the "raw" side of the plant to the "cooked" side of the plant.

A more appropriate audit plan for such a product would be to trace backward through the system. This strategy starts with the shipping end of the process and then works toward the receiving end. Trace backward audits minimize the possibility of cross-contamination of products with microbiological hazards. In addition, they allow the auditor the perspective of seeing the results of the process first.

The discovery method audits the functions in a random approach. This process has a disadvantage in that the auditor may get disoriented during the audit and may not audit all the critical activities. Therefore, utilizing this approach alone tends to become a hit-or-miss situation. However, coupling discovery with other approaches, such as the trace backward method, fortifies the audit methodology (the methods are more effective if they are used in combination rather than separately). In fact, a strong audit strategy for a food system is one that will use all the methods, ensuring the system is observed and evaluated from various angles.

The element approach focuses on auditing according to a specific standard or requirement. Examples include auditing to the five preliminary steps and seven principles of Codex HACCP or auditing to the elements of ISO 22000:2005.

Once the audit strategy has been decided, the plan for sampling and collecting data must be developed. After that, the audit strategy can be documented in the audit checklist. The checklist provides a focus for the actual execution of the audit. It has numerous functions, including:

- Ensuring that the audit is systematic and thorough and conducted against specific audit requirements
- Providing a record of the auditor's activities
- Defining the audit sampling plan
- Providing a document on which to record notes and objective evidence as the audit is being conducted

- Providing the basis for the audit report and corrective action reports
- Providing a time management tool for an efficient audit
- Serving as the framework for the closing meeting and audit report

Many clients have developed a generic checklist. When a generic checklist is used, it should be supplemented with time-specific or circumstance-specific items. For example, previous audits may have identified that an organization has minimal compliance with a specific PRP. As a result, the auditor may want to supplement the checklist with additional questions to ensure the potential problem has been properly addressed. Generic checklists also need to be supplemented with operation/facility-specific evaluators.

Traditionally, the checklist items are written as yes/no questions. During the course of the audit, the auditor must convert the checklist questions to open-ended questions. Checklists also may be developed as a series of statements, with the auditor required to give a grade such as fully meets, substantially meets, does not meet, failure, or critical failure. In addition, the auditor must develop a response when the audit element does not fully meet the stated requirements. Responses must be clear and concise.

When an assessment is required, a scoring rubric or standard needs to be developed to assist the auditor in scoring issues uniformly. Audit checklists should be flexible because the ultimate objective is to ensure an efficient and effective audit. Sliding scale checklists or any checklists on which auditors are expected to apply a grade require auditor standardization so scoring standards are applied uniformly across auditee departments and facilities by each member of the audit team.

Deviations from the original audit checklist may result because of a change in the schedule or because of observations made during the audit. When deviations occur, they can be noted on the checklist. Examples of checklist forms are shown in Figures 13.2 and 13.3. Figure 13.2 provides a general checklist form, while Figure 13.3 provides a detailed checklist that is used by National Marine Fisheries Service for its HACCP audits.

Audit Checklist

Audit number

Client

Subject — Checklist revision no.

Auditor — Page ___ of ___

Item no.	Ref. no.	Items audited/questions	Notes	Conform	
				Yes	No

Figure 13.2 Audit checklist (example).

Seafood Inspection Program
United States Department of Commerce
Systems Compliance Rating
(REV: June 15, 2005)

Name and Address of Facility Audited	Region
	Facility Number
	Phone Number
	CFN Number
Products Concerned	FAX Number
Auditors	E-Mail
Name and Title(s) of Accompanying Individual(s)	Dates of Audit and Report Delivery

Minor(m) Deficiency: A failure on the part of the system relative to facility sanitation which is not likely to reduce materially the facility's ability to meet acceptable sanitation requirements.

Major(M) Deficiency: A significant deviation from plan requirements, such that maintenance of safety, wholesomeness, or economic integrity is inhibited.

Serious(S) Deficiency: A severe deviation from plan requirements such that maintenance of safety, wholesomeness, or economic integrity is prevented; and, if the situation is allowed to continue, may result in unsafe, unwholesome, or misbranded product.

Critical(C) Deficiency: A hazardous deviation from plan requirements such that maintenance of safety, wholesomeness, or economic integrity is absent; will result in unsafe, unwholesome, or misbranded product.

	EXECUTIVE SUMMARY	m	M	S	C	P
1.0	MANAGEMENT CONTROLS AND RESPONSIBILITIES					
2.0	FOOD SAFETY					
3.0	SANITATION AND PREREQUISITE PROGRAMS					
4.0	QUALITY SYSTEM					
5.0	FOOD SECURITY					
	TOTAL					

1.0	MANAGEMENT CONTROLS AND RESPONSIBILITIES					
1.1.0	Management Responsibilities	m	M	S	C	P
1.1.1	Management commitment not properly implemented or communicated.					
1.1.2	Food safety policy not prepared or properly implemented.					
1.1.3	Food safety management system planning not properly performed.					
1.1.4	Responsibility or authority not properly defined or communicated.					
1.2.0	Food Safety Team	m	M	S	C	P
1.2.1	Food safety team leader not appointed.					
1.2.2	Food safety team leader does not report to top management.					
1.2.3	Food safety team is not interdisciplinary as applicable.					
1.3.0	Communication	m	M	S	C	P
1.3.1	Effective external communication not established, implemented, or maintained.					
1.3.2	Effective internal communication not established, implemented, or maintained.					
1.4.0	Emergency Preparedness and Response	m	M	S	C	P
1.4.1	Emergency response procedures not established, implemented, or maintained.					
1.5.0	Management Review	m	M	S	C	P
1.5.1	Management review not properly performed or documented.					
1.6.0	Resource Management	m	M	S	C	P
1.6.1	Necessary human resource competencies not identified.					
1.6.2	Personnel have not received documented training necessary for the proper function of the food system.					
1.6.3	Insufficient infrastructure to implement and maintain the food safety system.					
1.6.4	Work environment is not properly established, managed, or maintained relative to food safety.					
1.7.0	Continual Improvement	m	M	S	C	P
1.7.1	Continuous improvement activities not performed.					
2.0	FOOD SAFETY					
2.1.0	Hazard Analysis	m	M	S	C	P
2.1.1	Description of products, processes, and control measures not properly performed.					
2.1.2	Hazard analysis not properly performed.					
2.1.3	Hazard analysis not available.					

Figure 13.3 National Marine Fisheries Service audit checklist that uses a rating system.

2.2.0	HACCP Plan	m	M	S	C	P
2.2.1	No written HACCP plan when one is required.					
2.2.2	Plan is not location and/or fish species specific.					
2.2.3	Hazard is not listed in the plan.					
2.2.4	Hazard is not controlled.					
2.2.5	CCPs are not properly identified in the plan.					
2.2.6	Appropriate critical limit is not listed in the plan.					
2.2.7	Critical limits not followed.					
2.2.8	Monitoring procedure stated in plan is inadequate.					
2.2.9	Monitoring procedures not followed.					
2.2.10	Corrective action listed in plan is not appropriate.					
2.2.11	Corrective action not taken.					
2.2.12	Verification procedure stated in plan is inadequate.					
2.2.13	Verification procedures not followed.					
2.3.0	Control of Nonconformity	m	M	S	C	P
2.3.1	Traceability system inadequate.					
2.3.2	Improper handling of potentially unsafe products.					
2.3.3	Withdrawals or recalls not designed or implemented properly.					
2.4.0	Validation	m	M	S	C	P
2.4.1	Validation activities improperly performed.					
2.5.0	Records	m	M	S	C	P
2.5.1	Inadequate information on records. (Facility name and location, etc.)					
2.5.2	Record data is missing.					
2.5.3	Records are inaccurate.					
2.5.4	Records are not available for inspection.					
2.5.5	Documents or records are falsified.					
3.0	SANITATION AND PREREQUISITE PROGRAMS					
3.1.0	Sanitation Standard Operating Procedures and Prerequisite Programs	m	M	S	C	P
3.1.1	Sanitation Standard Operating procedures or prerequisite programs not present or effective.					
3.1.2	Sanitation standard operating procedures not followed.					
3.1.3	Sanitation not monitored.					
3.2.0	Safety of Process Water	m	M	S	C	P
3.2.1	Unsafe or unsanitary water supply.					
3.2.2	Water potability certificate not current.					
3.2.3	Self water treatment performed improperly.					
3.2.4	No protection against backflow, back-siphonage, or other sources of contamination.					
3.2.5	Inadequate supply of water and hot water.					
3.2.6	Ice not manufactured, handled, or used in a sanitary manner.					
3.3.0	Food Contact Surfaces	m	M	S	C	P
3.3.1	Equipment and utensils' design, construction, location, or materials cannot be readily cleaned or sanitized; does not preclude product adulteration or contamination.					
3.3.2	Equipment, primary packaging materials, and utensils not maintained in proper repair or removed when necessary. (Product contact surfaces)					
3.3.3	Product contact surfaces not cleaned and sanitized before use, after interruptions, or as necessary.					
3.3.4	Concentrations of cleaners and sanitizers are not effective or routinely checked.					
3.4.0	Prevention of Cross Contamination	m	M	S	C	P
3.4.1	Grounds condition can permit contamination to enter the facility.					
3.4.2	Facility					
3.4.2.1	Design, layout, or materials used cannot be readily cleaned or sanitized; does not preclude contamination. Insufficient lighting for the applicable operation.					
3.4.2.2	Insufficient separation by space or other means allows product to be adulterated or contaminated.					
3.4.3	Condition of roof, ceilings, walls, floors, or lighting not maintained; lights not protected.					
3.4.3.1	Areas directly affecting product or primary packaging material.					
3.4.3.2	Other.					
3.4.4	Cleaning methods permit adulteration or contamination.					
3.4.5	Finished product/primary packaging material not properly covered or protected.					
3.4.6	Equipment and utensils not maintained in proper repair or removed when necessary. (Non-product contact surfaces)					
3.4.7	Non-product contact surfaces not cleaned before use.					
3.4.8	Processing or food handling personnel do not maintain a high degree of personal cleanliness.					
3.4.9	Processing or food handling personnel do not take necessary precautions to prevent adulteration or contamination of food.					
3.5.0	Hand Washing, Hand Sanitizing, and Toilet Facilities	m	M	S	C	P
3.5.1	Hand washing and hand sanitizing stations not present or conveniently located.					
3.5.2	Improper disposal of toilet waster or sewage.					
3.5.3	Inadequate supplies. Absence of signs directing employees to wash their hands.					
3.5.4	Insufficient number of functional toilets.					
3.6.0	Protection from Adulteration	m	M	S	C	P
3.6.1	Condensation and other deleterious sources present.					
3.6.2	Adequate air exchange does not exist.					
3.7.0	Proper Labeling, Storage, and Use of Toxic Compounds	m	M	S	C	P

Figure 13.3 National Marine Fisheries Service audit checklist that uses a rating system *(continued)*.

3.7.1	Chemical(s) improperly used or handled.					
3.7.2	Chemical(s) improperly stored.					
3.7.3	Chemical(s) improperly labeled.					
3.7.4	MSDS sheets not available for all chemicals in use at the facility.					
3.8.0	**Control of Employee Health Conditions**	m	M	S	C	P
3.8.1	Facility management does not have in effect measures to restrict people with known disease from contaminating the product.					
3.9.0	**Exclusion of Pests**	m	M	S	C	P
3.9.1	Harborage and attractant areas present.					
3.9.2	Pest control measures not effective.					
3.9.2.1	Exclusion					
3.9.2.2	Extermination					
3.9.3	Improper disposal of processing waste.					
3.9.4	Inadequate housekeeping.					
3.9.5	No written pest control program.					
3.9.6	Pesticides not applied by a licensed individual.					
4.0	**QUALITY SYSTEM**					
4.1.0	**Management Responsibilities**	m	M	S	C	P
4.1.1	Management commitment not properly implemented or communicated.					
4.1.2	Food quality policy not prepared or properly implemented.					
4.1.3	Quality system planning not properly performed.					
4.1.4	Responsibility and authority not properly defined or communicated.					
4.2.0	**Quality Team**	m	M	S	C	P
4.2.1	Quality team leader not appointed.					
4.3.0	**Internal Communication**	m	M	S	C	P
4.3.1	Effective internal communication not established, implemented, or maintained.					
4.4.0	**Management Review**	m	M	S	C	P
4.4.1	Management review not properly performed or documented.					
4.5.0	**Resources Management**	m	M	S	C	P
4.5.1	Necessary human resource competencies not identified.					
4.5.2	Personnel have not received documented training necessary for the proper function of the quality system.					
4.5.3	Insufficient infrastructure to implement and maintain the food quality system					
4.6.0	**Quality Manual**	m	M	S	C	P
4.6.1	Quality manual is inadequate.					
4.6.2	Defect action plan is not adequate to control product quality characteristics.					
4.6.3	Defect action plan/quality manual not followed.					
4.7.0	**Product Requirements and Specifications**	m	M	S	C	P
4.7.1	Product characteristics not properly described including raw materials, ingredients, and end product.					
4.7.2	Intended use and reasonably expected handling of the product not properly considered.					
4.7.3	Product requirements not discussed and agreed with the customer.					
4.7.4	Labels and/or specifications are inadequate.					
4.7.5	Nonconforming product is improperly controlled.					
4.8.0	**Purchasing**	m	M	S	C	P
4.8.1	Evaluation, re-evaluation and selection criteria for suppliers is not established.					
4.8.2	Purchasing documents are not clear, reviewed, approved, or adequate.					
4.8.3	Verification of purchased product not properly performed or documented.					
4.8.4	Customer property not properly maintained or controlled.					
4.9.0	**Measurement, Analysis, and Improvement**	m	M	S	C	P
4.9.1	Customer satisfaction/dissatisfaction data not maintained or monitored.					
4.9.2	Internal audits not established or properly performed.					
4.9.3	Analysis of data and continuous improvement not properly performed with regard to the system.					
5.0	**FOOD SECURITY**					
5.1.0	**Management**	m	M	S	C	P
5.1.1	A comprehensive Food Security Plan has not been written, implemented, and periodically reviewed by the processor.					
5.2.0	**Human Element**	m	M	S	C	P
5.2.1	Access to plant or sensitive areas of the facility (by employees or visitors) is not sufficiently restricted to authorized personnel.					
5.2.2	Appropriate controls are not required of employees for gaining access to the facility.					
5.2.3	Hiring practices do not include a screening process.					
5.3.0	**Facility**	m	M	S	C	P
5.3.1	Facility, including outside premises, grounds, and perimeter, are not properly secure.					
5.4.0	**Operations**	m	M	S	C	P
5.4.1	Raw material suppliers are not subject to a documented approval/screening process.					
5.4.2	Supplier COCs or invoices do not address the subject of product origin and food security.					
5.4.3	Product integrity is not assured from time of shipping raw materials to processor through delivery of finished product to end-user.					

Figure 13.3 National Marine Fisheries Service audit checklist that uses a rating system *(continued)*.

The auditors may find it valuable to utilize additional audit tools while conducting an audit. These tools include:

- Corrective action request (CAR) (Figure 13.4)
- Comprehensive schedule (Figure 13.5)

Corrective Action Request

Organization audited:	CAR no.
	Audit no.
	Audit date

Food safety management system requirement

Document ____________ Rev ______ Sec/Par ______

Finding, problem, or potential problem:

Auditor ____________________ Date __________

Responsible manager ______________ Date __________

Review of magnitude of problem:

Response date ______________

Date completed:

Signature ____________________ Date __________

Corrective action:

Figure 13.4 CAR form (example).

Comprehensive schedule					
Date	Time	Topic	Auditee	Location	Notes

Figure 13.5 Comprehensive schedule form (example).

- Finding and comment log
- Corrective action log
- Audit worksheet

The HACCP audit may start with a desk audit. This phase of the audit may be conducted off-site and prior to the start of the formal audit process. The function of the desk audit is to review appropriate documentation prior to the on-site audit and ensure that the documents describe a food safety management system that meets regulatory, customer, and any internationally designated standards. The initial desk audit has several advantages, including:

- Utilizes the on-site audit time more effectively
- Reviews specific documents for compliance with standards, regulations, and customer requirements
- Develops a better understanding of the auditee's processes to help focus the audit
- Enables the audit team leader to build a team with the appropriate expertise
- Enables the audit team leader to prepare appropriate and useful audit work documents

A desk audit may be conducted for first-party HACCP audits. This process can help the auditor prepare for the on-site audit. In addition, a desk audit can reduce the time the auditor spends on the site.

Typical documents that may be reviewed prior to the on-site audit include the following:

- Food safety manual
- Food safety goals and objectives
- Food safety policy
- Hazard analysis report

- Risk assessment report
- HACCP plan and PRP plans
- Past corrective action reports
- Past audit reports
- End product list
- Ingredient and raw product list
- Training documents

These documents need to comply with requirements such as food laws and regulations, customer specifications, the appropriate standards (such as ISO 22000), and the Codex Alimentarius Commission publications applicable to the operation.

CONDUCTING THE AUDIT

The objective of the audit is to determine:

- Whether the food safety management system meets the stated requirements
- Whether controls are effective and correctly implemented to assure the production of safe food

Therefore, audits focus on fact finding rather than fault finding. Assume that the auditee is innocent until proven guilty. In general, the auditor should not take the attitude that things are going wrong. It is not "gotcha." This tactic tends to put a lot of stress on the organization, and the employees' actions may not reflect actual day-to-day actions.

The on-site portion of the audit should be conducted using the formal management system audit process, which consists of the following elements:

- Opening meeting
- Collection of data
- Analysis of results
- Closing meeting

Opening Meeting

The opening meeting sets the stage and tone for the data gathering phase of the audit. For external audits, these meetings are formal. For internal audits, the meetings tend to be less formal since many of the issues need not be covered with the auditors. However, it is recommended that internal audits be launched with an opening meeting.

The opening meeting clarifies the roles and responsibilities of both the audit team and the auditee. The opening meeting is attended by representatives of the auditee and the audit team. At the opening meeting, the lead auditor is responsible for doing the following:

- Thanking the host
- Introducing the audit team
- Stating the purpose and scope of the audit
- Describing the documents used to prepare the audit plan
- Describing any changes to the audit plan
- Explaining the audit methods and techniques that will be used in the audit
- Possibly referring to previous audits and corrective actions
- Possibly asking specific questions with regard to any documentation that has been reviewed
- Possibly identifying areas of interest that the auditee, client, or lead auditor want audited
- Possibly presenting an audit schedule, if the auditee does not have a schedule
- Verifying that the auditee has communicated the audit plan to employees, security departments, and any applicable unions
- Communicating the following: the expected time of the closing meeting, how nonconformances will be handled, and the expected time of delivery of the audit report
- Verifying the logistics for the audit

The lead auditor also is responsible for ensuring that an attendance log and minutes of the opening meeting are kept.

The auditee's management is responsible for the following at the opening meeting:

- Reviewing any applicable corporate procedures, including safety and environmental policies
- Introducing the auditee's management team
- Reviewing policies that govern proprietary rights
- Stating the availability of various amenities such as meeting rooms, and areas where the auditors can secure personal and corporate equipment and documents, phones, copiers, computers, mobile devices, restrooms, eating facilities, etc.
- Identifying the individual who will represent the auditee during the audit and any escorts
- Requesting any clarifications if the report from the lead auditor does not meet the auditee's expectations

- Verifying that the auditor understands the company's good manufacturing practices (GMPs) and personnel safety requirements for visitors.
- Presenting an overview of the plant: parking, telephone, restrooms, and so on

An audit escort should be provided for each member of the audit team. The escort plays a critical role in the conduct of the audit. The escort becomes the representative of the auditee to the audit team. As a result, this person is responsible for the following:

- Obtaining requested supplies and documents for the auditor
- Acting as an audit observer for the auditee's management
- Acting as a guide for the auditor
- Introducing the auditor to individuals he/she wishes to interview
- Ensuring that the auditor follows all applicable corporate policies and procedures, including safety rules
- Serving as an interface between the auditor and the site
- Communicating to the organization any changes in the audit schedule
- Confirming or denying nonconformances

If the auditors have not previously visited the facility, a tour may be appropriate prior to the start of the data gathering phase of the audit. The tour needs to be worked into the schedule, and the lead auditor should ensure that the tour is not excessive. If the lead auditor believes the tour is excessive and is not of value to the audit process, he/she should call a halt to the tour and insist that the data gathering portion of the audit begin.

Collection of Data

Once the opening meeting is completed, the data collection phase of the audit begins. The data gathering phase of the audit is the most time-consuming portion of the audit. Data can be gathered using several techniques, including:

- Document and record review
- Interviews
- Observation of work activities
- Physical examination of product

In addition, the audit team will review any written plans the auditee did not provide in advance of the on-site audit.

Sometimes these activities are done as individual activities, and other times the activities are done in combination. The auditor must gather objective evidence that the food safety management system is effectively implemented to assure the production of safe food. Objective evidence is information that is proven to be true and verifiable.

Document and record review is a critical part of the audit. Documents specify what should be done, and records provide objective evidence of what activities have been conducted. The document and record review links the auditee's past performance to current performance. This provides a systematic and complete approach to auditing so auditors can focus on issues including:

- Assessing areas where repeat problems have occurred in audit reports
- Conducting reviews to ensure the documents are adequate to meet the objectives of the food safety management system

The organization needs to make the appropriate, current documents available to auditors to help them execute assigned responsibilities.

Records should be accurate, complete, and written in ink, and they should appear as if created during the normal course of business. If a datum entry mistake was made, the correct value should be entered on the record. The mistake should be crossed out with a single line so the original value can be read. The person who made the correction should initial and date the record. If the audit team finds that records are incomplete, incorrect, or conflicting, the team must resolve these discrepancies.

Much emphasis is placed on record review in a food safety system. Records are often considered evidence in a court of law when the situation warrants. However, record review is only a part of the HACCP audit. The record review process provides evidence of whether the record-keeping system is acceptable and can be trusted.

For example, if the review and analysis of records is acceptable, it provides evidence that the product produced during a specific time period is safe. But if other evidence tells the auditor that a particular lot was determined not to be safe, the record-keeping system is suspect. If several record-keeping incidents are uncovered during the audit, there is evidence that even the lots that may have been deemed acceptable are now suspect since the information in the entire HACCP system cannot be trusted.

A review of all records generated since the last HACCP audit often is not feasible and would be too time-consuming. The auditor should develop a mechanism for sampling the records.

When sampling records, the lead auditor must determine not only the sample size but also how the sample of records will be chosen. For example, for a sample size of 30 records over the past three months, the auditor could start with record number three and skip every five records until the sample size of 30 is reached. Most important, however, is that when reviewing the records in such a sample, the auditor must determine the significance of any deficiency noted, how many such deficiencies must be present to accept or reject the system, and what action to take when deficiencies are found. Not having the firm's address on a record is a deficiency, but not one that is significant or severe. However, a verified record illustrating that a critical limit was exceeded without corrective action is highly significant. When sampling records for a food system, it is often useful to sample by days of production, as this can give the auditor a feel for the activities of a particular production day or shift. Prior to the audit, the lead auditor should request samples of records, develop the audit standard, and review it with the audit team.

Interviews provide the auditor with answers to specific questions. During this part of the audit, the auditor should determine how the organization is responding to new and emerging hazards and whether the organization is using newer intervention technologies to respond to the hazards.

Care must be used in evaluating interview data. Any verbal response should be considered hearsay unless the response is corroborated by other audit evidence. Verbal evidence becomes stronger when the audit team hears similar responses from individuals in other parts of the organization or if the auditor can corroborate the interview with records. Other issues that the auditor must consider in evaluating interview data are the bias of the interviewee and whether the person could have any hidden agendas.

Audits are stressful events for the auditee. During an interview, the auditor must first put interviewees at ease (often by explaining what is being done). Next, the auditor needs to find out what interviewees are doing as part of their responsibilities in ensuring the production of safe food. This can be done by asking open-ended questions, asking them to show the records they keep, or asking them to show what they do as part of their job responsibilities. Try to develop a conversation rather than grilling or interrogating to determine if operators and technicians know their roles and responsibilities.

Sometimes a person being audited will freeze during the interview. The auditor must determine whether the interviewee does not understand the question or cannot answer the question. The auditor needs to ensure he/she is effectively communicating to the interviewee, especially line personnel. The auditor may rephrase a question to ensure that the interviewee understands the intent of the question.

Most individuals are proud of what they do and are willing to explain their jobs and their responsibilities. It is the auditor's responsibility to determine if the explanation meets the requirements of the job. If the interview process is done correctly, individuals will tell you why they are doing something—even if it is wrong. The auditor should be aware that a person being interviewed may make suggestions for improving the food safety system. When this is done, the auditor must evaluate the suggestions and determine whether they should be incorporated into the audit report. (Use discretion—hidden agenda issues may be presented as seemingly valuable suggestions, and auditors must be careful to avoid becoming pawns in an interviewee's scheme to gain some job advantage.)

The auditor must always be careful that the escort is not intentionally steering the auditor toward specific individuals to be interviewed and/or away from other individuals. Select interviewees randomly. Auditors need to be alert to escort posturing. Some escorts may place themselves in a position where they can influence interviewees' statements or responses. Care must be taken in selecting individuals with production responsibilities to interview. This is to ensure that the audit process does not adversely affect the normal manufacturing process.

A large part of the HACCP audit is observation of work practices, equipment, and facilities. Auditors always must remember that audits are disruptive to the normal work process. Individuals are never at ease when someone is observing them. Care must be taken during this part of the process because in many operations, additional personnel in the work space may disrupt the flow of product or limit the space needed to conduct a test and record the results.

To conduct this part of the process, auditors need to be well trained in observation techniques and understand the process. The lead auditor should confirm that each audit team member has the appropriate criteria to determine if operators are executing their responsibilities as written. During this phase of the audit, the auditor needs to establish the following:

- Is the product being made in accordance with the HACCP plan?
- If the product is being made in accordance with the HACCP plan, are the PRPs effective in maintaining the processing environment to produce safe food?
- Do the employees know their roles and responsibilities in following the HACCP and PRP plans?
- Do operators know their roles and responsibilities?
- Are the food safety documents and records being properly used and maintained?
- Are there gaps in the HACCP system between the execution and auditee documents and either the NACMCF standard, the Codex Alimentarius Commission standard, or other food safety management system standards?
- Are the PRPs integrated to ensure the proper environment is present to manufacture safe food?
- Do the operations being conducted appear choreographed?

The auditor should look for subtle signs that the PRPs are deteriorating. One strategy is to ask a person from management how he/she knows that a specific PRP is operating effectively.

Time management is critical to ensure an effective audit. Auditors must use their discretion when choosing areas for in-depth investigation or observation.

If on the surface it appears that the cleaning and sanitizing processes are being followed, then the HACCP auditor may want to concentrate on assessing the difficult-to-clean areas. For example, when auditing a fish processing plant, the employees may do a good job in cleaning and sanitizing the cutting table; however, they may neglect to do a good job cleaning and sanitizing pumps, since the pumps require disassembly prior to cleaning and sanitizing. In this case, during the evaluation of the PRPs the HACCP auditor may want to inspect a sample of the pumps to determine if the plant is properly following the cleaning and sanitizing program. Experience has shown that if the easy places to clean are dirty, the entire plant will be dirty.

The auditor should observe plant operations over all manufacturing shifts. Many food manufacturing systems operate with two production shifts and one cleanup shift. It is critical that the audit team assess the system on both manufacturing shifts and on the sanitation shift. Sanitation is a part of the PRPs and a critical component of the food safety system. If the plant operates multiple days between sanitation activities, the audit should be scheduled so the sanitation process can be observed.

PRPs that include GMP compliance programs can be a weak spot in the food safety management system since, psychologically, PRPs do not have the same level of criticality as the HACCP plan. If a critical limit is violated, unacceptable product is produced from a food safety perspective. Typically, a single incident of a slight deviation in a PRP is unlikely to cause a food safety incident. Therefore, if this type of deviation is observed during a HACCP audit, it may be reported as a minor issue rather than a major issue.

As an example, during the audit of a warehouse, the auditor sees chipped paint on a safety pole that is used to protect a doorjamb. The auditor determines that the location of the chips will not contaminate the ingredients or the finished product. This can be considered a minor deviation in the PRP, since the processing facilities are not maintained. However, this minor deviation may cause the auditor to look further into other problems with regard to facilities maintenance.

Slight deviations in the PRP are not minor in the long term. Food safety problems can occur if there is a breakdown of a PRP. The FDA reported that between 1999 and 2003 more than 81% of Class I and II recalls were attributed to GMP-related problems.[4]

Experience has shown that if these systems are not properly maintained and controlled, they can rapidly deteriorate and cause a food safety incident. If the HACCP auditor uncovers several deviations in PRPs, the combination of the deviations may be recorded as a major issue. In addition, the auditor may want to assess the company's commitment to maintaining a strong PRP.

The auditor's responsibility is not to uncover problems that are outside the scope of the audit. However, if this type of problem is discovered, the auditor should not ignore it. The auditor's reaction should depend on the severity of the problem and the effect the problem will have on the integrity of the food safety management system. For example, say an auditor discovers that an employee is not wearing earplugs in an area that requires the use of earplugs. This is not a deviation within the scope of auditing a food safety management system. However, the auditor may want to bring this incident to the attention of the organization's management. In addition, the auditor may make additional observations to determine whether employees are using other personal protective equipment properly as part of the food safety management system, such as hairnets, beard nets, sleeve protectors, and gloves.

The auditor should keep the auditee's management informed of any significant problem as it is uncovered. This can be done by communicating these issues through the escort or during briefing meetings with the auditee's management. The goal is to prevent surprises at the closing meeting.

If a HACCP audit takes place over several days, it is common to have briefing meetings with the auditee's management and caucus meetings for the audit team. The briefing meetings are designed to update the organization's management on the progress of the audit, resolve any issues, and discuss any problems that might have been observed during the audit. The team caucus meetings keep the team informed about developments in the audit, allow for audit schedule revisions, enable auditors to compare audit evidence, and help the team start to reach consensus on the audit results.

Analysis of Results

The analysis of data is the final step in the data collection process. It is the evaluation of audit results to determine if the food safety system conforms to the stated goals and objectives. In addition, it determines if the food safety management system is effective and efficient in preventing a food safety incident. Typically, the analysis is done throughout the audit. However, the audit team normally meets prior to the closing meeting to conduct a final analysis of the facility, classify the evidence, and develop the agenda for the closing meeting.

All the facts need to be discussed in the final analysis meeting. Each finding must be developed into a clear and concise statement of the problem, linked to audit requirements. In addition, each finding should be supported by objective evidence. Once a finding is identified, the auditor may elect to follow an audit trail, which can be useful to the company in determining the root cause of the finding. For example, unclean tables in a processing room can be the result of poor training, poor personnel practice, or poor supervision. The auditor may elect to follow any or all of the potential leads. The auditor needs to determine if the findings are major or minor in effect. Once this is done, the nonconformance is drafted.

The lead auditor needs to be aware of both corporate and regulatory policies that govern serious and critical findings. Food processing companies may set internal critical limits tighter than the regulatory policies. This allows the food processor to take corrective action before a regulatory policy is exceeded while allowing the company to continue to ship product. For example, federal regulations to control the histamine levels in fresh tuna require that the tuna temperature can exceed 41°F for a maximum of eight hours. A HACCP plan may state that the tuna temperature can exceed 38°F for a maximum of eight hours. If an auditor discovers that tuna has been held at a temperature of 39°F for eight hours, this is a violation of the HACCP plan but not a violation of the regulatory requirement. Thus, the incident may not be a food safety problem; however, it is an audit finding since the company has violated its internal standards. As a result, the company needs to take appropriate corrective action. Furthermore, if the process continues to deteriorate, there might be a breakdown of the system that could lead to a food safety incident.

Minor issues should not be overlooked during the analysis phase, since deterioration of a minor problem can lead to a future food safety incident. It is also appropriate (and important) for the auditor to identify areas where the organization should be commended.

Objective evidence is used to determine the degree of conformance to the food safety objectives and standard. Nonconformances must be verifiable and traceable. The analysis will be used as a basis to develop audit findings and report nonconformances. It also serves as the basis for the closing meeting and the audit report.

Audit findings are drafted during the closing meeting. Individuals should not be named in or connected with findings; use position names or phraseology such as "the operator at…" It may be appropriate to name an individual where the individual has specific knowledge of the situation. If the client is responsible for approving and issuing the actual nonconformance, any nonconformance documents that are presented to the auditee during the closing meeting should be marked as draft or not presented at all.

Closeout dates for findings may be a fixed date, or they may be set through negotiations between the auditor and the auditee and based on the criticality of the finding and the actual time required to complete the tasks properly. One way to deal with findings that may take time to complete is to set a date by which the company will deliver a plan that will be used to remove the root cause of the finding. In future audits, the auditor can review the plan and determine if the company is meeting the milestones that are stated in the plan. Corrections, at least on a temporary basis, should be performed as soon as possible.

For example, the seafood inspection program of the National Marine Fisheries Service sets a limit of two weeks to correct major or serious deficiencies. The agency may allow for up to one year to correct a structural problem. However, in both cases production of food is only allowed if the immediate deficiency does not allow the production of food containing a health hazard.

Once the analysis is completed, the audit team should be able to develop a unified response that describes the adequacy and effectiveness of the food safety system. It is best if the final analysis is a consensus of the audit team. However, if there is disagreement among the team members, the audit team leader has the responsibility of resolving the disagreement, and the auditors have the responsibility of supporting the team leader during the closing meeting. The primary strategy to reduce the number of disagreements in developing and reporting findings is to stick to the facts and the objective evidence.

Closing Meeting

The closing meeting is held at the conclusion of the formal audit. The closing meeting is used to present the audit results to senior management and ensure that management clearly understands the results.

Typically, the closing meeting is attended by the same individuals who attended the opening meeting. Sometimes higher levels of management attend the closing meeting. The following is a typical agenda for a closing meeting:

- Thank the host.
- Reaffirm the purpose, scope, and objectives of the audit.
- Present an audit summary and describe the results.
- Describe how the results were prioritized.
- Describe the details of the audits, including presenting the findings, concerns, and any commendations.
- Verify acknowledgment of the findings.
- State that the audit does not uncover everything. There may be additional nonconformances.
- Verify the follow-up procedures and resolution of findings.
- Inform the auditee of when they will receive the audit report.

The lead auditor is responsible for ensuring that there is an attendance list and that minutes are kept.

The closing meeting is not the forum for detailed debate over a nonconformance or a finding. However, if the auditee can produce information or evidence that would affect the finding, the audit team must consider this prior to completion of the audit report. All findings should be presented at the closing meeting. The auditee should not find any surprises, either positive or negative, in the audit report.

The lead auditor is responsible for ensuring: (1) that the nonconformance or finding is clearly defined; (2) that the process is described for effective corrective action in both determining the cause and implementing an appropriate solution; and (3) that there is an emphasis on the timely resolution of the corrective actions and for ascertaining the need, if any, for a follow-up audit and the departments that would be affected by the follow-up audit. If the nonconformances pose a significant risk, there will be a special follow-up. During the closing meeting, the escort may help explain the nonconformance or finding.

Audit Report, Follow-up, and Closeout

The audit report provides formal written documentation of the audit results. The audit report has a number of customers, including the client and the auditee. The ultimate customers of the report are the line personnel in the organization. These individuals are responsible for the day-to-day activities that ensure the production of safe food.

The audit report should be sent to the auditee within a mutually agreed-upon time frame. The sooner the audit report is delivered the better. For external audits, the time frame for responding to audit findings starts when the auditor leaves the company. There tends to be a reduction in the urgency of the corrective actions and an increase in miscommunications when audit communiqués are delayed. The audit report formally documents the audit and initiates the corrective action and system improvement. In addition, the report guides auditees in their subsequent decisions and actions. The audit report contains the following elements:

- Date the report is issued
- Date of the audit
- Organization audited
- Purpose, scope, and objective of the audit
- Details on the itinerary and timetable
- Identification of the audit team
- Identification of the auditee's representatives
- Identification of the audit criteria and standards
- Distribution list
- Executive summary

- Record of the audit
  - Opening meeting
    - Summary of the meeting
    - Attendance list
  - Observations
    - Supporting evidence associated with the finding
    - Comments
    - Areas of conformance
    - Areas of nonconformance
    - Areas of concern
    - Commendations
  - Closing meeting
    - Summary of the meeting
    - Attendance list
    - Positive points observed
    - Review of nonconformances
    - Discussion of recommendations
    - Designated follow-up
    - General observations
    - Best practices and commendations
    - Auditee comments about nonconformances if appropriate and significant
    - Follow-up
- Follow-up and closeout requirements

As part of an internal HACCP audit, the audit report may also include the audit plan and the audit checklist. The audit checklist can provide a means to document audit observations and objective evidence. If the audit checklist is not used to document the auditor's notes, then the auditor's notes should be included in the audit report. The audit report also may contain the audit's working papers such as the detailed audit schedule and the completed checklist.

The HACCP audit must go beyond ensuring that the organization is in compliance with the stated food safety policies and procedures. It needs to determine the status of the food safety management system and whether this system can achieve the food safety goals and requirements in an efficient and effective manner. Food safety issues are not fixed. Over time, new biological, chemical, and physical hazards are identified. In addition, new intervention techniques are developed that can control existing and new hazards. The HACCP audit also must focus on these issues. This ensures that the food safety management system will remain effective in the future.

For first-party audits, recommendations can be provided in an appendix or separate report that is clearly identified as recommendations. For second- and third-party audits, recommendations should not be made. Many times the auditor can provide recommendations on how to remediate the finding. However, extreme care must be exercised. In making recommendations, it is possible for the auditor to take partial ownership of the auditee's system. This should never occur, even in first-party audits. The internal auditor should focus on auditing. It is possible that the internal auditor within a small company will also be part of the team that addresses the findings. When this is the case, the person who served as the auditor should not act like an auditor but as a member of the improvement team.

A critical aspect of audit closeout is the appropriate resolution of corrective actions. This entails the organization taking proper action to contain the short-term problems and to develop and implement strategies to prevent reoccurrence. During the process, the auditee's management must do the following:

- Set priorities for the CARs.
- Identify the individuals responsible for resolving the findings.
- Identify the underlying causes and trigger events.
- Determine if the problem can occur in other areas of the organization.
- Develop a solution for the nonconformance.
- Develop a plan and a schedule to correct the deficiency.
- Implement the plan.
- Implement any new control measures.
- Verify the effectiveness of the corrective action.
- Develop preventive action.
- Implement preventive action.
- Verify effectiveness of the preventive action.

Upon completion, the auditee should prepare a report. This provides formal documentation and objective evidence that the CAR has been properly addressed and appropriate actions were taken to eliminate the root causes of the finding.

Once the organization has taken appropriate actions to prevent a reoccurrence, the auditor can close out the findings and CARs. The verification activities do not mean that a second on-site audit needs to take place. The actual action that takes place depends on the type of nonconformance, the nature of the corrective action, and the client's verification requirements. It may be sufficient for the auditor to review evidence showing that the corrective action has been executed and is effective in preventing reoccurrence. In this case, the auditor can verify the adequacy of the corrective action at the next audit. If the nonconformance is severe, the auditor may recommend a follow-up audit. The client has the responsibility to determine if a follow-up audit is necessary. The scope of the follow-up audit is to ensure effective implementation of the corrective action and development of a preventive action. In either case, the review must focus on whether the auditee's corrective actions were implemented and whether they were appropriate to prevent reoccurrence of the nonconformance or if preventive actions are indicated.

If the corrective action is neither implemented nor effective, the auditor should first reevaluate the entire situation. If the reevaluation indicates that further action must be taken, the auditor can do one of the following:

- Issue a new corrective action that has been escalated; for example, upgrading a minor finding to a major finding.
- Issue a corrective action on the absence of an effective corrective action process.
- Escalate the corrective action by taking it to higher-level management or emphasizing its importance to the client.

If audit findings indicate frequent recurring corrective actions or corrective actions that have been issued on the same symptom, the auditor should suspect that the corrective action process is not functioning in an effective manner.

Some organizations will track the progress of corrective actions; an example of a corrective action tracking form is shown in Figure 13.4.

CLOSURE

Audit closure takes place when all corrective actions for an audit have been closed (or implemented and verified as agreed). The lead auditor should communicate to the auditee indicating that all corrective actions have been completed and that the audit is closed.

NOTES

1. Centers for Disease Control, "Update: Multistate Outbreak of *Escherichia coli O157:H7* Infections from Hamburgers—Western United States, 1992–1993" (Atlanta, 1993), http://www.cdc.gov/mmwr/preview/mmwrhtml/00020219.htm (last accessed June 11, 2005).
2. Peter Feng, "*Escherichia coli Serotype O157:H7*: Novel Vehicles of Infection and Emergence of Phenotypic Variants" (Washington, DC: U.S. Food and Drug Administration, 1995).
3. U.S. Food and Drug Administration, BBB—*Escherichia coli O157:H7* (EHEC) (Washington, DC, 2012).
4. U.S. Food and Drug Administration, "Food GMP Modernization Working Group: Report Summarizing Food Recalls, 1999–2003" (Washington, DC: CFSAN/Office of Scientific Analysis and Support, 2004).

SUGGESTED READING

Russell, J. P., ed. *The ASQ Auditing Handbook,* 4th ed. Milwaukee, WI: ASQ Quality Press, 2013.

———. *Internal Auditing Pocket Guide: Preparing, Performing, Reporting and Follow-up,* 2nd ed. Milwaukee, WI: ASQ Quality Press, 2007.

———. *The Process Auditing and Techniques Guide,* 2nd ed. Milwaukee, WI: ASQ Quality Press, 2010.

Sayle, A. J. *Management Audits,* 3rd ed. Brighton, MI: Allan Sayle Associates, 1997.

Chapter 14

Auditing Process and Auditor Competencies

AUDITOR COMPETENCIES

The capabilities and competencies of a HACCP auditor are keys to a successful audit. HACCP auditors need to be knowledgeable in a number of areas, including:

- Techniques of management system audits
- Food safety microbiology
- Food processing fundamentals
- Food safety issues
- Food safety management systems, including the current principles of HACCP and prerequisite programs (PRPs)
- Applicable food laws and regulations of the food processing system being audited
- Application of auditing principles to food processing systems
- Risk and hazard assessment

HACCP auditing involves evaluating the risk assessment process the company uses to identify biological, chemical, and physical hazards; hazard analysis; and studies used to validate the critical control points (CCPs).

The HACCP auditor in the food processing sector needs to be aware of emerging problems and appropriate interventions to assess the auditee's HACCP system properly. Auditors always must be aware of the gravity associated with food processing operations, since failures of these systems can lead to death. Thus, the auditor needs to be able to assess the food safety management system for current effectiveness as well as continued effectiveness in light of the emergence of new pathogens.

An effective auditor requires a number of qualities to conduct a successful audit. The person must be able to interview and interact with a wide variety of personnel in a food processing operation. The auditor must be equally at ease when discussing food safety systems with plant management and with line personnel.

During an audit, there may be times when an auditee representative or employee becomes defensive or even angry or upset. The auditor must have the skills to remain calm and help the person being interviewed regain emotional control so the interview can resume. One technique is declaring a "time out" to allow the interviewee to regain composure. The auditor must have strong observational skills to develop an assessment of how good food safety management practices fit with the various processing techniques in use at the auditee facility. Finally, the auditor needs to have strong oral and written communication skills to effectively communicate audit results.

Business conduct and local customs vary from location to location. Auditors need to be aware of the local customs and practices, especially when auditing internationally. What may be considered appropriate behavior in one culture may be considered inappropriate behavior in a different culture. The auditor needs to be cognizant of what is appropriate behavior within the auditee's organization. The auditor needs to behave in a manner that does not compromise his/her reputation or impede or compromise the audit process.

When auditing an organization, the auditor must either be fluent in the agreed-upon language or have available at all times an interpreter with the appropriate skills to interpret technical information. The auditing organization should provide the interpreter.

In multicultural situations, some of the employees of the auditee may have weak language skills in the agreed-upon language. It is necessary that the auditor determine if language or literacy barriers will interfere with the audit process. If so, the auditor may need an interpreter. If an employee has some language skills in the language of the auditor, the auditor may achieve the desired results by asking simple questions and requesting the employee to demonstrate assigned work tasks.

Auditors need to recognize language hurdles and compensate by speaking more slowly, using correct grammar, and avoiding contractions, slang, and ambiguity. Speaking loudly does not make a word easier to understand.

ETHICS

Ethical behavior is based on a conclusion of whether an action is right or wrong. Ethical behavior often is defined by moral principles and guidance found in a person's culture, society, laws, regulations, or professional conduct dictates. The fundamental concept with regard to auditing ethics is that the decisions and the auditor must be honest and impartial. In auditing a system, auditors must conduct themselves in a professional manner, using objectivity and honesty as their guiding principles. The American Society for Quality (ASQ) has developed a strong code of ethics for certified individuals, including those holding such certifications as Certified Quality Auditor and Certified HACCP Auditor. Many other organizations, corporations, and governments have developed such codes to define acceptable behavior. All have similar principles regarding conflict of interest, confidentiality, proprietary information, and the handling of unacceptable situations. Adopting and adhering to such codes of ethics enables the auditor to transcend various personalities, styles, and temperaments; it assists in maintaining a high standard of performance and conduct. The ASQ Code of Ethics is presented in Appendix F.

Conflict of Interest

The auditor must identify potential conflicts of interest and take appropriate action when or before they arise. The credibility of the audit process is highly dependent on the credibility of the auditor. Conflict of interest situations include but are not limited to previous employment of the auditor by the auditee, a financial interest in the auditee's business affairs, assisting in the development of the system being audited, an application for employment held by the auditee, and so on. Whenever a conflict of interest exists, the auditor must make the situation known to the lead auditor, client, auditee, and other appropriate parties for resolution.

Proprietary Information and Confidentiality

Much of the information encountered by auditors of a food safety system is highly proprietary or confidential. The premature or inappropriate release of this information could lead to devastating financial effects. Formulas, trade secrets, processing methods, and so on typically are found in the performance of food safety audits. Auditors must be diligent in maintaining the security of this information. Auditors often are requested to sign a confidentiality agreement that binds them legally in such situations prior to the commencement of the audit. Care should be taken when discussing proprietary information to make certain security is maintained.

In some management audits, the auditee may refuse access to the auditor to part of the manufacturing process for reasons of proprietary information or confidentiality. In these cases, the auditor may elect to "audit around" that part of the process; the auditor will assess the inputs and outputs of the process without assessing the actual process itself. Unfortunately, in a HACCP audit, unlike other audits such as quality system audits, it is very difficult to audit around a room or situation since failure to actually review the entire process could lead to severe consequences.

LIABILITY

Liability is a legal term that indicates the responsibility an entity (individual, company, corporation, society) possesses in any particular situation. Issues of liability are becoming increasingly important in auditing circles and in certification systems. One only has to read the headlines to understand the liabilities being faced by large auditing firms and their clients. Auditors must accept some liability for decisions indicating acceptability of a food safety system. It must be remembered that the public, not just the auditee, could be relying on the auditor's decisions. Thus, auditors must be able to stick to the facts found in their report and in their notes. Further, findings must be properly supported, and records must be kept a sufficient time beyond the shelf life of the product produced.

ILLEGAL OR UNSAFE ACTIVITIES

If unethical or potentially illegal activities, including violations of food processing regulations, are detected, the auditor must verify the situation and inform the lead auditor. It is the lead auditor's responsibility to bring these issues to the attention of the client and/or auditee. The auditor must bring these issues to the auditee's attention when food safety could be compromised. It is management's responsibility to take the appropriate corrective action. If management is sponsoring these activities, then the auditor should seek legal counsel to resolve these issues. Specific certification schemes will often provide guidance in such cases.

If the unethical activities involve a potential violation of the ASQ Code of Ethics, the activities should be reported to the local ASQ section for investigation and possible reporting to the ASQ Ethics Committee.

INTERNATIONAL REGULATIONS AND THIRD-PARTY AUDITING SCHEMES

As the food supply became global, so did the various regulations, systems, and schemes that developed. World trade is focused primarily on the codes and guidelines from Codex Alimentarius (http://www.fao.org/fao-who-codexalimentarius/en/) where government trade and international trade in various food products is agreed upon and published. As auditors gain more experience, they will undoubtedly encounter a firm trading globally that must be aware of the Codex practices. As such, it is incumbent on the auditor to research and be familiar with any Codex Codes of Practice on the product form being audited. As global trade increases, governments will promulgate regulations to accommodate the various products entering their country. The European Union and the United States both have such regulations in place. For example, recently the FDA published the Foreign Supplier Verification Program rule, which requires most firms to "provide adequate assurances that their foreign suppliers are producing food in compliance with processes and procedures that provide at least the same level of public health protection" as those required under the Food, Drug and Cosmetic Act. (https://www.registrarcorp.com/fda-food/food-safety/fsvp/?utm_source=bing&utm_medium=cpc&utm_term=fda%20fsvp&utm_content=1229254207724900&utm_campaign=386357627&matchtype=e&device=c&msclkid=049edaf66bed1b308510851779cd9f2a)

Private schemes have also increased in the last two decades to address the increased desires of the supplier to address consumer expectations above and beyond governmental standards in food safety. These schemes all have their own published standards. Some are based on the ISO 22000 family of standards. Some focus on a particular subject, process, concern, or industry segment. Following is a partial list of these schemes and standards:

- British Retail Consortium (BRC) standards
- Primus
- Safe Quality Food (SQF) Food Safety Code
- FSSC 22000 standards
- Global Good Agriculture Standards (G.A.P.)
- Global Aquaculture Alliance

No auditor is expected to be expert or conversant in all such schemes and regulations. Also, an auditor will ultimately be exposed to situations in which a single firm is audited against a number of these standards and by various audit groups. In fact, some larger firms employ a staff simply to deal with the various systems and auditors they face in a given year. What is important is for the auditor to understand that a firm may have more than one system it must address simply because of its various buyers. This does not mean there are different food safety systems; rather, it means there are different ways food safety is expected to be addressed by customers. Understanding the existence of the various schemes and regulations, and becoming aware of those that apply to the situation at hand, is imperative for the successful auditor to study during the planning phase of the audit.

SUGGESTED READING

International Organization for Standardization. ISO 22000:2005, *Food safety management systems—Requirements for any organization in the food chain*. Geneva: ISO, 2005.

Russell, J. P., ed. *The ASQ Auditing Handbook*, 4th ed. Milwaukee, WI: ASQ Quality Press, 2013.

———. *Internal Auditing Pocket Guide: Preparing, Performing, Reporting and Follow-up,* 2nd ed. Milwaukee, WI: ASQ Quality Press, 2007.

———. *The Process Auditing and Techniques Guide,* 2nd ed. Milwaukee, WI: ASQ Quality Press, 2010.

Sayle, A. J. *Management Audits,* 3rd ed. Brighton, MI: Allan Sayle Associates, 1997.

Chapter 15

Quality Tools and Techniques

MEASUREMENT OF DATA

Quality and product safety data can be classified as either attribute data or variable data. Attribute data are measurements that are counted. When using an attribute measure, the analyst determines whether an individual item possesses or does not possess a specific characteristic, such as whether salmonella is either present or absent in a sample. The characteristic being measured can be either a positive or a negative attribute. Variable data typically are physical, chemical, or biological measurements such as length, weight, temperature, and pH. Measurement data are variable data.

Data normally are described using two statistics:

- The central tendency
- The amount of dispersion

Central tendency is reported as the mean, median, or mode. The mean is calculated as the arithmetic average. The mean is normally the default method of reporting the central tendency. The median is the middle number of a sample set. It is used when the data are highly skewed. The mode is the most common number or numbers. It is used when reporting a set of data using a histogram. In addition, it is the preferred method of reporting data when the data have several peaks or are multimodal.

The most common method of reporting the amount of dispersion is the standard deviation. Some individuals may refer to the standard deviation as sigma (Σ). The following formula is used to calculate the standard deviation for variable data:

$$s = \sqrt{\frac{\sum \left(x - \bar{x}\right)^2}{n - 1}}$$

The standard deviation has some unique properties with regard to normally or near normally distributed data. These properties can be summarized in the following manner:

Approximately 67% of the values will fall between $\bar{x} \pm 1$ standard deviation.

Approximately 95% of the values will fall between $\bar{x} \pm 2$ standard deviations.

Approximately 99.73% of the values will fall between $\bar{x} \pm 3$ standard deviations.

For small data sets (less than 10 values), the amount of dispersion may be reported as the range or the maximum number minus the minimum number.

Prior to statistical analysis, microbiological data may be transformed using a log10 transformation. This is done to ensure that the microbiological data are relatively normally distributed.

FLOW DIAGRAMS

The flow diagram or flowchart depicts all the steps of a process. Standard symbols are used to identify the process steps, decision steps, and documentation steps. Flow diagrams break complex processes into meaningful subcomponents for analysis. Figure 15.1 shows a typical flow diagram.

There are two major types of flow diagrams: classical industrial engineering flow diagrams and schematic diagrams. Classical industrial engineering flow diagrams show the following:

- Sequence and interaction of process steps
- Location of outsourced or subcontracted processes
- Location for rework and recycling
- Areas with potential for process delays
- Location where raw ingredients, packaging material, and intermediate products enter the process
- Location where intermediate products, waste, and finished products leave the process

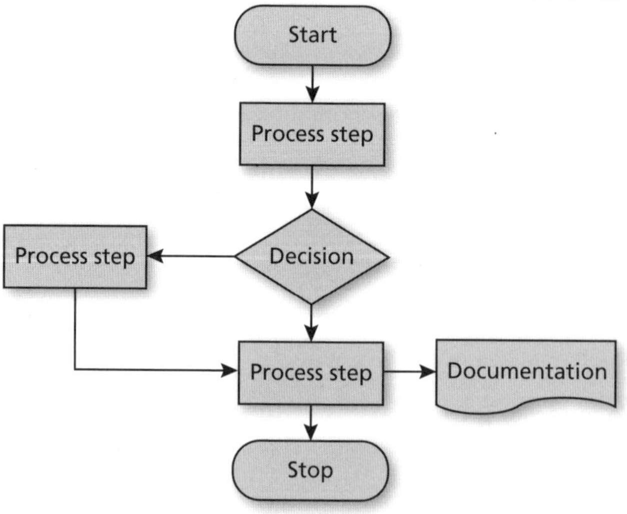

Figure 15.1 Flow diagram (flowchart).

Source: © J. G. Surak, used with permission of the author.

Schematic diagrams show the following:

- Flow of products
- Flow of employees
- Flow of rework
- Location of high-risk and low-risk areas
- Location of storage and distribution areas
- Flow of waste
- Flow of water
- Flow of air and other utilities

PARETO ANALYSIS

Pareto diagrams (Figure 15.2) rank the relative importance of different nonconformities or nonconforming items. This analysis tool is used because many complex problems such as the root cause of failures or nonconforming items can have multiple causal factors. Typically, the problems are measured against the frequency of occurrence. The Pareto analysis is helpful in identifying and justifying which defect must be resolved first or where the most likely cause of a problem is.

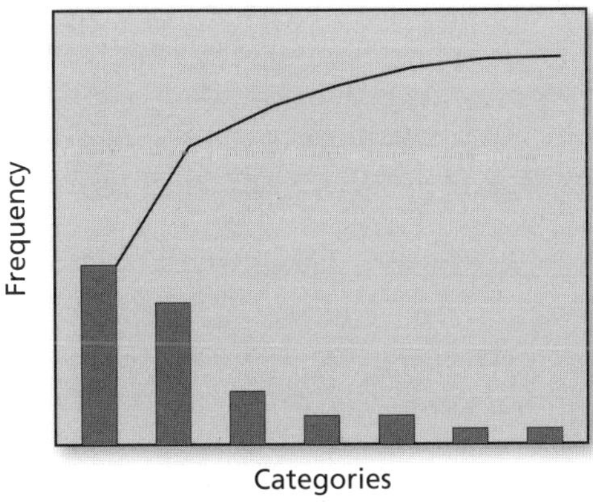

Figure 15.2 Pareto diagram.

Source: © J. G. Surak, used with permission of the author.

CAUSE AND EFFECT DIAGRAMS

The cause and effect diagram (Figure 15.3) is also known as the fishbone diagram or Ishikawa diagram. The cause and effect analysis can be used as part of the root cause problem-solving process to identify potential links between a cause of a problem and the effect of the problem. A cause and effect analysis segments problems in a logical and convenient order by separating the primary, secondary, and tertiary factors.

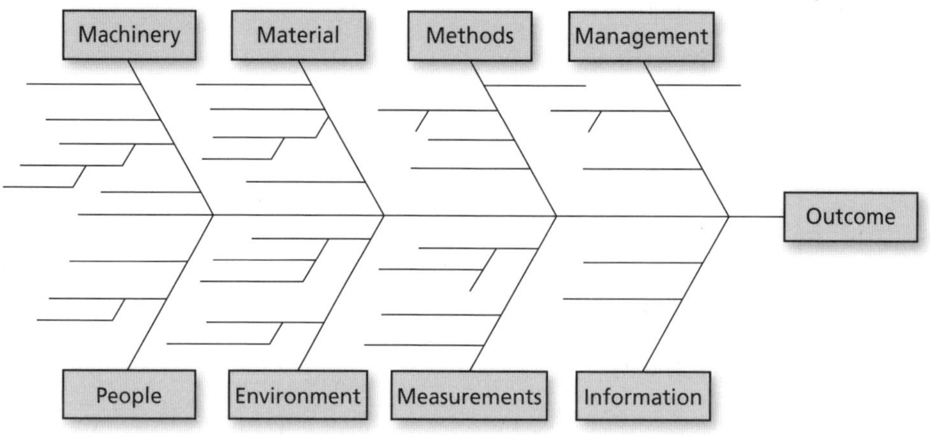

Figure 15.3 Cause and effect diagram.

Source: © J. G. Surak, used with permission of the author.

HISTOGRAMS

A histogram (Figure 15.4) is a graphical display of tabulated frequencies that shows which proportion of cases fall into each of several categories. The categories are usually specified as nonoverlapping intervals (or bins) of some variable. The categories (bars) must be adjacent. Since the data are presented as individual values, specifications can be placed on the histogram. Therefore, the diagram can be used to show the relationship of the mean, size of variation, and shape of variation to both the upper and lower product and process specifications. Histograms can be used to determine if multiple distribution is occurring in a process.

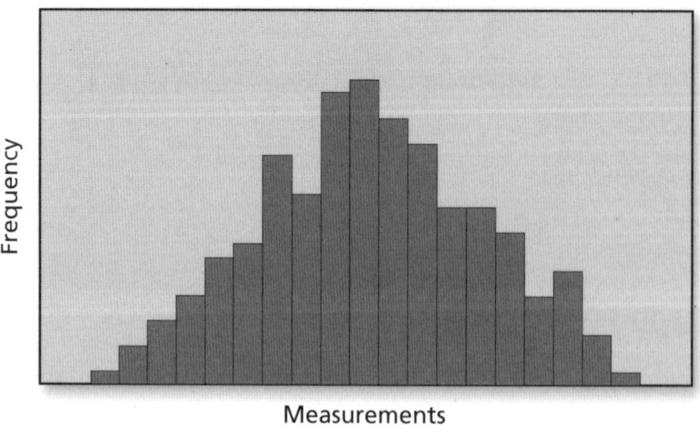

Figure 15.4 Histogram.

Source: © J. G. Surak, used with permission of the author.

SCATTER DIAGRAMS

A scatter diagram or scatter plot (Figure 15.5) is a graph used to display and compare two or more sets of related numerical data by displaying several data points, each having a coordinate on a horizontal and a vertical axis. For example, a scatter diagram could be developed to correlate the addition of an acidulant to the pH of a batch. Scatter diagrams can be used to show the extent of variation between two variables. Scatter diagrams cannot prove a cause and effect relationship. This proof must be found using other problem-solving techniques such as cause and effect analysis.

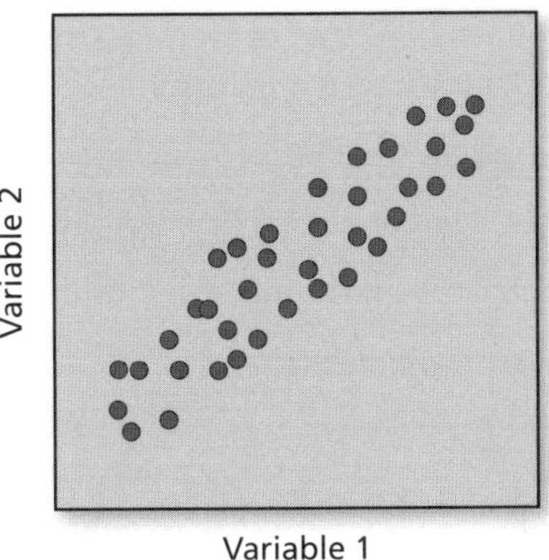

Figure 15.5 Scatter diagram.

Source: © J. G. Surak, used with permission of the author.

CHECK SHEETS

Check sheets are records that make it easy to compile data so they can be readily used in data analysis. The auditor checklist is a check sheet. Check sheets should be in the simplest form to help the individual collect the data. Check sheets also can be used to ensure that a specific set of steps in a process is followed. This type of check sheet will require some sort of signature to record the event. If check sheets are to be a valid record, they must be completed as the work is being done. Auditors use check sheets to ensure that evaluators are assessed. Auditors should take care to avoid becoming a tool of the check sheet. The auditor's objective is to record findings, not to limit observations to the checklist points.

CONTROL CHARTS

Control charts or Shewhart charts are used to determine if process variation is stable. The control chart is a statistical tool intended to assess the nature of variation in a process. Control charts are used to facilitate management of food safety and quality issues.

When process variation is stable, it is possible to derive upper and lower control chart limits. These limits are set at the mean plus or minus three standard deviations of the standard deviation of the average.

Variable control charts (Figure 15.6) are composed of two charts; the top portion is the plot of the average values, and the bottom portion is a chart of either the range or the moving range.

The control chart is used to determine if there has been a change in the process that affects the process variation. This is accomplished by classifying the observed variation as due to either common causes of variation or special causes of variation.

Common cause variation is random (or predictable) variation. A process with only common cause variation is considered to be in statistical control. Common cause variation is characterized by:

- Phenomena constantly active within the system
- Predictable variation
- Variation within a historical experience base
- Lack of significance in individual high or low values

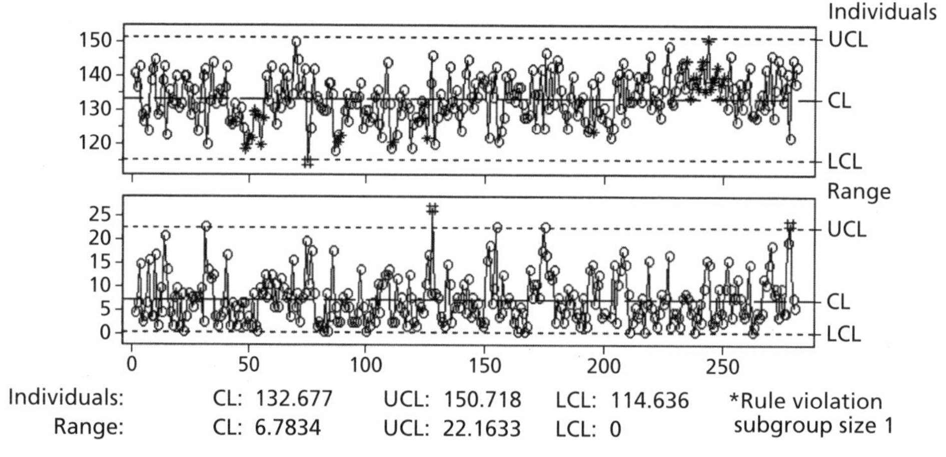

Figure 15.6 Variable control chart.

Source: © J. G. Surak, used with permission of the author.

Figure 15.7 shows an example of a control chart where there are only common causes of variation. The outcomes of a roulette wheel are a good example of common cause variation. Common cause variation is the noise within the system.

Special cause variation, or assignable cause variation, always arrives as a surprise. It is the signal within the system. Special cause variation is characterized by:

- New, unanticipated, emergent, or previously neglected phenomena within the system
- Inherently unpredictable variation

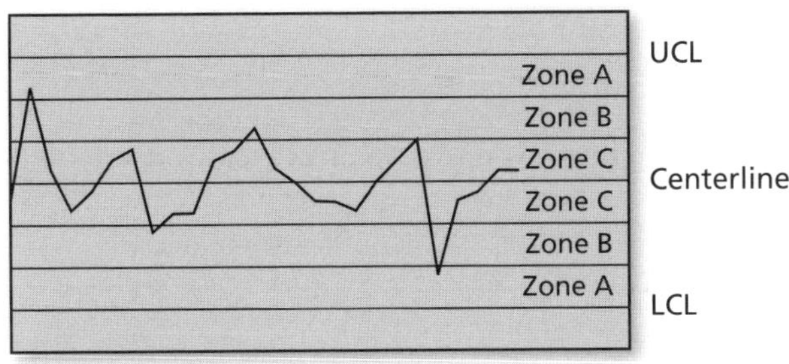

Figure 15.7 Control chart showing only common causes of variation.

Source: © J. G. Surak, used with permission of the author.

- Variation outside the historical experience base
- Evidence of some inherent change in the system or our knowledge of it

When a process shows a special cause of variation, the organization should take action to identify the cause of variation and eliminate or prevent the reoccurrence of the cause of variation. A control chart can be divided into three zones on each side of the central line. Zone C encompasses the values between the central line (mean or $\bar{x}$) and the mean plus (or minus) one standard deviation of the mean ($s_{\bar{x}}$). Zone B encompasses the values from the mean plus (or minus) $1s_{\bar{x}}$ to the mean plus (or minus) $2s_{\bar{x}}$. Zone A encompasses the mean plus (or minus) $2s_{\bar{x}}$ to the mean plus (or minus) $3s_{\bar{x}}$. The zones are used to help in the analysis of control charts. A process is said to be affected by a special cause of variation when one of the following observations is made on the control chart:

- One point exceeds the upper or lower control chart limit.
- Two out of three consecutive points are in zone A.
- Four out of five consecutive points are in zone A or zone B.
- Eight consecutive points are on one side of the central line.

Figures 15.8–15.11 show examples of control charts that display special causes of variation.

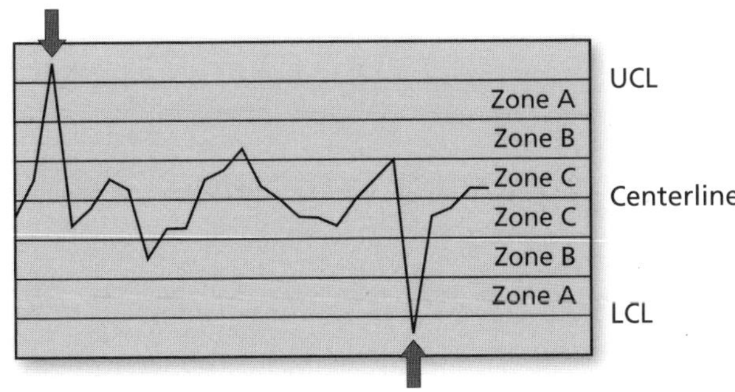

Figure 15.8 Control chart showing a special cause of variation where one point exceeds the upper or lower control chart limit.

Source: © J. G. Surak, used with permission of the author.

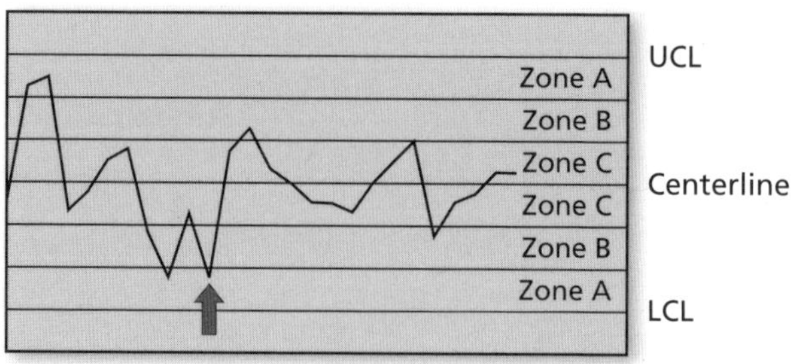

Figure 15.9 Control chart showing a special cause of variation where two out of three consecutive points are in zone A.

Source: © J. G. Surak, used with permission of the author.

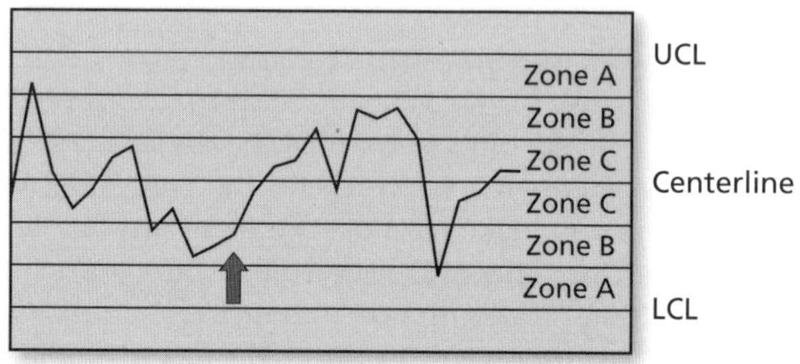

Figure 15.10 Control chart showing a special cause of variation where four out of five consecutive points are in zone B or zone A.

Source: © J. G. Surak, used with permission of the author.

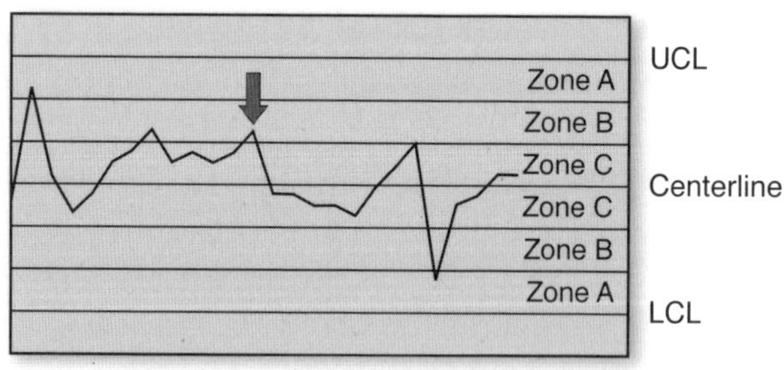

Figure 15.11 Control chart showing a special cause of variation where eight consecutive points are on one side of the central line.

Source: © J. G. Surak, used with permission of the author.

Variable data are plotted on mean–range ($\bar{x}$ and R) control charts or individual moving range (xMR) control charts to measure the variation of the data.

Attribute data classically are charted on one of four types of control charts. The type selected is dependent on the type of attribute data and whether the sample size is constant or variable.

np—Chart of the number of nonconforming units; the subgroup size is constant.

p—Chart of the proportion of nonconforming units; the subgroup size is variable.

c—Chart of the number of nonconformities; the subgroup size is constant.

u—Chart of the proportion of nonconformities; the subgroup size is variable.

CAPABILITY INDICES

Capability indices determine the relationship between the process or product specifications and the process variability. There are two primary capability indices: C_p and C_{pk}. C_p is the ratio of the specification tolerance and six times the standard deviation of the individual values. The following formula is used to calculate C_p:

$$C_p = \frac{\text{Upper specification} - \text{Lower specification}}{6 * s}$$

C_{pk} is a complex measure. It measures the relationship between the process mean and the specification and the relationship between the specification tolerance and the process variability. The following formulas are used to calculate C_{pk}:

$$C_{pk} = \left[\text{Minimum}\left(\frac{\text{Upper specification} - \bar{x}}{3 * s}\right) \text{ or } \left(\frac{\bar{x} - \text{Lower specification}}{3 * s}\right)\right]$$

Some Six Sigma programs attempt to differentiate between short-term process capability and long-term process capability. The difference between the two is the method used to calculate the standard deviation. Some quality professionals do not support the use of short-term process capability because one of the objectives of process capability indices is to show the relationship between the variation the customer will observe and the specification limits. When this concept is used, the only variation the customer will see is the long-term variation. When an auditor reviews process capability data, he/she should ask for the conditions under which the data were collected, and the equations used to calculate the process capability indices.

CONTINUOUS IMPROVEMENT

Continuous improvement is a central pillar of both the quality and food safety movements. The objective of continuous improvement is to remove the sources of variation or the sources of nonconformances. This normally is done using root cause analysis. The objective of root cause analysis is to get beyond the systems and identify and remove the underlying reason for the problem. Effective root cause analysis consists of the following steps:

- Determine the potential causes of a problem.
- Select the most probable underlying cause or causes.
- Develop a hypothesis of the most probable cause or causes.

- Test the hypothesis.
- Separate the root cause from any contributing causes.
- Develop a plan to eliminate the root cause or reduce its effects to an acceptable level.
- Implement the new procedure or control.
- Verify the effectiveness of the process changes.
- Report the results.

A number of the continuous improvement tools are used throughout the root cause analysis. Two problem-solving models have been used to aid individuals in root cause analysis: the plan–do–check–act (PDCA) cycle and define–measure–analyze–improve–control (DMAIC).

The PDCA cycle is the traditional method for managing the continuous improvement process. It describes how the scientific method can be applied to solve operational problems. The PDCA cycle consists of the following parts:

Plan	Plan a study or a test. Develop hypotheses. Select the best hypothesis.
Do	Conduct a study to test the hypothesis. Collect data during the study.
Check	Check or analyze the results of the study. Determine whether the study proved or disproved the hypothesis.
Act	Act on the results. If the study was successful, roll out the solution. If the study was not successful, develop another hypothesis.

DMAIC is the continuous improvement method of Six Sigma. When DMAIC is applied to continuous improvement, it is very similar to the PDCA cycle. It presents the scientific method using a slightly different set of descriptors and adds a fifth step, control. This last step is not a new step; it is implied in the PDCA cycle. DMAIC as applied to continuous improvement consists of the following parts:

Define	Define the problem and potential causes of the problem. Define potential solutions and select the best solution.
Measure	Measure the effect of the solution on the process in solving the root cause of the problem.
Analyze	Analyze the data and determine if the solution is effective.
Improve	Improve the process.
Control	Develop controls to maintain the gains.

CORRECTIVE ACTION/PREVENTIVE ACTION

CAPA is an acronym that stands for corrective action/preventive action. Corrective actions are reactions to problems or failures. They are actions taken to remove the cause of a nonconformance. Corrective action in the food industry is taken against the product and is the barrier that prevents a hazard from causing consumer illness or injury. Corrective actions have two parts: (1) the immediate actions taken to contain the problem that is caused by the nonconformance; and (2) actions taken to ensure the nonconformance does not reoccur. Issues that trigger corrective actions include audit findings and nonconformances, customer complaints, and process control reports that indicate either a nonstable process or a noncapable process.

Preventive actions are actions taken to eliminate or reduce the cause of a potential nonconformance. Issues that trigger preventive actions include adverse trends from data analysis or increased risks from risk analysis. The objective is to take action before a problem is realized. Root-cause problem-solving provides a tool to identify and eliminate the cause of the nonconformity or the potential cause of a nonconformity.

There is debate over whether preventive actions exist within the food safety management system. The group that states that preventive actions do not exist justifies its position on the basis that HACCP itself is a preventive action program. Others state that preventive actions are part of the food safety management system. If one looks at the structure of ISO 22000:2005, preventive actions are not formally mentioned. However, the standard requires that actions continually be taken to identify potential sources of hazards and prevent their occurrence.

SAMPLING

Sampling of a population must be done when it is impossible to conduct a 100% inspection of the entire population. The critical aspect of sampling is the collection of an appropriate sample. The objective of the sample is to collect data that are representative of the entire population.

Some auditors state that statistical sampling techniques are essential for appropriate audit results, while others state that it is not necessary to look at a statistically significant sample to determine if there is a problem with a management system. Many times, a proper statistical sample cannot be obtained for a number of reasons, including that it is difficult to obtain a truly random sample or the required sample size is too large. It should be noted that if the sample size is too small, one cannot have any statistical confidence in the results of the inspection. In any case, the auditor needs to take care in selecting a sample so the data represent appropriate issues within the food safety management system. There are several issues that must be addressed before a sample can be collected. These include the type of sample and the size of the sample, what is being sampled, and for what reason. Sampling an end product prior to its leaving the facility may require a different sampling plan than sampling a record-keeping system. Figure 15.12 presents a record and end-product sampling procedure that is used by the National Marine and Seafood Inspection.

Record Review

The objective of the review of the record-keeping system is to determine if the records are creditable and can be used as objective evidence as part of the audit process. The sample size is 12 days of records, randomly chosen, and all records are examined using the following criteria.

Criteria

Significant deviation:

- Missing entries for measurements or readings
- Calculation errors that indicate a higher safety or quality level
- Values changed without justification or initials
- Values on record do not agree with auditor's evidence from other sources
- Any deviation that would have a significant effect on the safety, wholesomeness, labeling, or quality of the final product

Minor deviation:

- Dates, addresses, or signatures missing
- Missing calculations, such as averages, that in themselves do not affect the acceptance of the product
- Any deviation that, although listed as required on the record, does not have a significant effect on the safety, wholesomeness, labeling, or quality of the final product

Evaluation

Accept the creditability of the record-keeping process if:

- No significant deviations are found
- Minor deviations are five or less

Reject the creditability of the record-keeping process if:

- Two significant deviations are found or
- Minor deviations are more than eight in number

Action if rejected:

- Consider a "Serious" deviation for "Records are inaccurate" if no other evidence exists. If it can be shown that the specific system failure did result in at least one lot of noncomplying product (through end item examination of product outside the facility), and significant deviations were found, consider a "Critical" deviation for "Records are inaccurate." If less than 12 days production, sample all days.

End Product Evaluation

During the system audit, the auditor evaluates no more than three lots of product, based on the definition of lot listed in the facility's HACCP plan. The method of discovery sampling is used, where the sample purpose is to locate one adverse factor or deviation. If any deviation is found, minor or significant, the auditor is to investigate the deviation found until it can be determined what significance it holds and the scope of the deviation. Once the root cause is found, the auditor will make an assessment of its significance or severity. If it is found that two of the lots do not meet compliance requirements, this would be considered a "Serious" deviation under "Records are inaccurate." If all three lots show noncompliance, this would be considered a "Critical" deviation under "Records are inaccurate."

If three lots are not available, including lots under production, and the lots show noncompliance, only consider the "Serious" deviation.

Figure 15.12 Example of a record sampling and assessment system and end-product sampling procedure.

Source: Seafood Inspection Program of the U.S. National Marine and Seafood Agency.

The type of sample concerns whether the population will be subdivided prior to sampling. In *random sampling*, all objects in the population have an equal chance of being selected for the sample. The random sample is taken to learn about the entire population. Many times it is not possible to select an actual random sample. Other times it is not desirable.

Stratified sampling divides the population into several strata or groups. Within each strata there is a greater amount of homogeneity within the sample. The sampling is designed to ensure that a sample is taken of each strata regardless of the size of each strata. Statistically valid comments can be made about each group but not about the population as a whole. An example of stratified sampling is developing strata based on shifts.

A third type of sampling is *block sampling*. The population is divided into rational blocks, and the sample is selected from those blocks. An example is to divide a population by blocks of days, such as January 1–15, January 16–31, and so on. The blocking technique is an excellent sampling technique when looking for a root cause of a problem. Statistically valid comments can be made about each block but not about the population as a whole.

Another sampling technique is *judgmental sampling*. This may be done if the auditor has prior knowledge of the population. Normally, samples are selected to determine if a problem is occurring or if there are high-risk areas that must be addressed in the scope of the audit. This sampling technique can be done if the probability that the problem will occur is relatively low.

One of the big issues is what constitutes a proper sample size. Since auditing focuses on determining whether a food safety management system conforms to a requirement, the auditor will be working primarily with attribute samples. Most auditors use ANSI/ASQ Z1.4-2008—*Sampling Procedures and Tables for Inspection by Attributes*. This standard is a revision of MIL-STD 105E—*Sampling Procedures and Tables for Inspection by Attributes*. MIL-STD 105E was withdrawn as a sampling standard in 1989. Auditors must always use the most recent version of any standard.

Table 15.1 provides the definitions used in acceptance sampling.

Table 15.1 Definitions in acceptance sampling.

Term	Definition
Acceptance quality level (AQL)	The quality level that is the worst tolerable product average when a continuing series of lots is submitted for acceptance sampling.
Producer risk (α risk)	The probability of nonacceptance when the quality level has a value stated by the acceptance sampling plan as acceptable.
Limiting quality level (LQL)	The quality level that for the purpose of acceptance sampling inspection is the limit of an unsatisfactory process average when a continuing series of lots is considered.
Consumer risk (β risk)	The probability of acceptance when the quality level has a value stated by the acceptance sampling plan as unacceptable.
Acceptance number (AC)	The largest number of nonconformities or nonconforming items found in a sample by acceptance sampling inspection by attributes that permits the acceptance of the lot as given in the acceptance sampling plan
Operating characteristic (OC) curve	A curve showing the relationship between the probability of acceptance of product and the incoming quality level for a given acceptance sampling plan

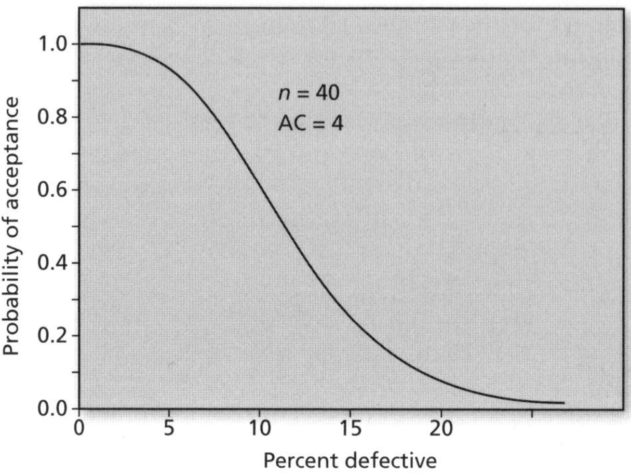

Figure 15.13 OC curve.

Source: © J. G. Surak, used with permission of the author.

The operating characteristic (OC) curve is generated for a specific sample size, a specific acceptance number, and a probable level of the fraction of nonconformances. It is a plot of the probability of acceptance versus the actual fraction of nonconformances. The OC curve can be used to determine the producer risk and the consumer risk. Figure 15.13 depicts an OC curve for a sample size of 40 with an acceptance number of 4. If the producer risk is set at 0.95 and the consumer risk is set at 0.10, the acceptance quality level (AQL) is 0.02 or 2%, and the limiting quality level (LQL) is 0.08 or 8%. For example, say the HACCP auditor uses this sampling plan and examines 100 documents. The auditor finds four nonconforming documents. The auditor does not know the actual number of nonconformities. Statistics calculate the most likely range of the actual number of defects. This range is 2%–8%. If the actual number of nonconforming documents is 6.8%, there is a probability of 0.1 that the auditor will say there is no problem. If the actual number of nonconforming documents is 3%, there is a probability of 0.8 that the auditor will say there is no problem. Whenever a sampling plan is used, the auditor needs to refer to the OC curve to determine if the producer risk and the consumer risk are acceptable.

Part V

Food Safety and Quality in the Food Processing Industry

Chapter 16

The Food Industry in General

FOOD COMPONENTS

Food is a complex biological system. The principal components of food products are cellular tissues from plants or animals or other products of animal origin such as milk and eggs. These commodities are combined with chemicals to form food. Chemicals are added to a food product for a number of reasons: to fortify with nutrients, to improve flavors, or to enhance other sensory aspects. Chemicals also can inhibit spoilage, including microbial spoilage.

Plant tissue used in food products can come from roots, stems, leaves, nuts, or fruits. This means that plant tissue can come in contact with a wide variety of potentially hazardous agents, including pollutants and intentionally applied toxic chemicals (pesticides). However, the primary hazards on plant material are microorganisms originating in the soil.

Muscle, the primary source of animal tissue, is traditionally classified in the following manner:

- *Meat.* Sourced primarily from domesticated cattle, swine, lambs, and goats
- *Poultry.* Sourced primarily from chicken, turkey, and squab
- *Fish.* Includes fin fish (both freshwater and saltwater), shrimp, crab, and molluscan shellfish

Animal tissue and products of animal origin can present special foodborne disease hazards because animals can be infected with zoonotic microorganisms, or microorganisms that can cause disease in both humans and animals. Even animals that are disease-free can harbor or carry microorganisms that cause foodborne disease in humans.

SOURCES AND TYPES OF FOOD HAZARDS

Food products are in a constant state of change. These changes can be caused by natural mechanisms or by humans during food processing operations. Changes start on the farm and continue through food processing operations. In addition, a finished food product is not immune to changes.

Changes to food can be beneficial or detrimental. Once a plant is harvested or an animal is slaughtered, the natural defense mechanism of the plant or animal is compromised, and the tissue starts to decompose. This decomposition can be caused by internal or external sources. Internal decomposition is autocatalytic and enzymatic in origin. External decomposition is caused by some external living system such as microorganisms, other animals, chemical reactions, or physical actions.

On the farm, beneficial changes include the maturing of vegetables, fruits, or animals. Another example of a beneficial change is the hydro-cooling and packaging of fresh-cut vegetables for the fresh produce market. Detrimental changes observed in animals include diseases and tumors. Detrimental changes to plant tissue include bruises, tumors or galls, and infections.

A number of microorganisms play an essential role in the production of fermented foods such as cheese, yogurt, pickles, sauerkraut, some types of sausage, and alcoholic beverages. Other microorganisms are detrimental to human health. For example, *Aspergillus flavus,* a mold that produces mycotoxins that can cause tumors in mammals, can infect plant material and cause a human health hazard.

Chemical changes may be enzymatic or nonenzymatic and may cause beneficial or nonbeneficial changes to food. Beneficial chemical changes include the development of a brown crust during the baking of bread. The bread crust turns brown as a result of the Maillard reaction—a reaction between reducing sugars and the amino groups of amino acids, peptides, and proteins. A classical undesirable chemical reaction is the formation of nitrosamines, a carcinogen, during the improper curing of fermented meats or sausage.

Physical changes include size reduction. Size reduction can add value to food products by cutting plant or animal tissue into useable sizes and by separating edible portions of animal or plant tissue from inedible portions. However, if the size reduction process is not controlled properly, it can increase the rate of autolysis of the tissue by disrupting an excess number of cells.

Commercial food processors expect that ingredients entering the processing plant, including farm-based commodities, will meet proper specifications. This minimizes the need for incoming inspection and sorting. Many food processors require farmers to apply the principles of HACCP to the production of agricultural products. Chapter 17 describes the application of HACCP to the meat and poultry industry, while Chapter 18 discusses the application of HACCP to the seafood industry. The dairy industry is discussed in Chapter 19. Chapter 20 describes how farmers and food processors can reduce hazards in some food components by applying HACCP principles to fresh fruit and vegetable production. Finally, Chapter 21 details the relationship of HACCP to the retail and food service industries. In all these settings, the primary changes in food products typically start with the changes that occur with harvesting or slaughter and continue through food processing, packaging, storage, and delivery.

Food processing is designed to enhance beneficial changes and to reduce the rate of detrimental changes. Converting raw ingredients into a tasty food product that can be served as a meal is an example of a beneficial change. Processed food can undergo two types of detrimental changes:

- Spoilage that makes the food inedible. The following four factors affect the spoilage rates of food: time, temperature, oxygen, and water. The spoilage of food can be catalyzed by biological, chemical (enzymatic or nonenzymatic), or physical mechanisms.

- Contamination of the food with a biological, chemical, or physical hazard. Biological hazards are the primary cause of food safety hazards. HACCP provides a mechanism to ensure that all hazards are properly addressed and controlled.

Food processing cannot prevent the deterioration of food, but it can retard the spoilage rate. Even if the product is commercially sterilized and packaged in a hermetically sealed package, nonenzymatic chemical changes can occur during the shelf life of the product and render it aesthetically unpleasing.

The following food processing steps are used to retard the deterioration of food: drying, heating, cooling (which includes both refrigeration and freezing), fermenting, irradiating, and packaging. Packaging is used to control the amount of oxygen that contacts the food product or to prevent contact from occurring. In addition, packaging can protect the product from physical abuse, contamination by microorganisms or vermin, or human tampering.

Microbial Growth in Food Products

Food processors can control microorganisms by controlling the physical properties of food through chemical or physical methods.

Microorganisms require water for growth. Water can be present in two forms in food: free water and bound water. Water that is present in food may or may not be available to microorganisms for microbial growth. Food technologists use water activity (A_w) as an estimate of the amount of water available for microbial growth. A_w is the ratio of the vapor pressure of the food to the vapor pressure of pure water. As A_w decreases, the amount of water available for microbial growth declines. *C. botulinum* does not grow in food with an A_w of less than 0.91. In general, bacterial pathogens do not grow in food products that have an A_w of less than 0.85. The lower limit for yeast growth is 0.7 to 0.75 and for mold growth is 0.6. A_w can be lowered in foods by adding ingredients such as salt or sugar. In addition, the water content of foods can be reduced by evaporation or drying.

The hydrogen ion concentration, or pH, also affects the growth of microorganisms. *C. botulinum* can grow in foods when the pH is greater than 4.8. Therefore, foods with a pH of less than 4.5 are classified as high-acid foods, and foods that have a pH of greater than 4.5 are classified as low-acid foods. Low-acid canned foods must undergo a thermal process to destroy *C. botulinum* spores. If the food product has a pH of less than 4.5, the thermal process is less severe because it needs to kill only the vegetative pathogens.

Various chemicals can be used to either retard the growth of or kill microorganisms. These chemicals can be divided into two major types: chemicals that can be added to food products or packages and chemicals that cannot be added to food products or packages. In addition to food-grade acids, which are used to reduce the food's pH, other antimicrobial agents include benzoic acid, sorbic acid, propionic acid, nisin, natamycin, nitrate, nitrite, sulfite, and sulfur dioxide. Compounds such as hydrogen peroxide are used as a chemical sterilant in aseptic packaging systems.

Biological Hazards

Microorganisms, the primary cause of biological hazards in food, are present throughout the environment. Microbial actions can be beneficial, innocuous, or detrimental to human health. Most types of microorganisms are innocuous to humans. However, those that are harmful to human health can cause death. HACCP auditors are encouraged to study a standard food microbiology text or the FDA's *Bad Bug Book*[1] for details on the growth of pathogens and the incidence of implicated foods. Table A.1 (Appendix A) lists some of the microorganisms that can cause biological hazards in foods.

To sustain their growth, microorganisms require environmental conditions similar to those required by humans. Key elements for growth include time, optimum temperature, nutrients, and water. Some microorganisms require oxygen for growth, while others occur only under anaerobic conditions. For example, *Clostridium botulinum* is a strict anaerobe that produces the deadly toxin that causes botulism. In contrast, *Listeria monocytogenes,* which has caused foodborne disease outbreaks in prepared meats, cheeses, and fresh vegetables, is an aerobic organism.

Since food is a natural source of nutrients and water for microorganisms, food processors use the following processes to control the growth and level of microorganisms.

Heat

Depending on the actual temperature of the product, heat has several effects on microorganisms. At temperatures just above the optimal temperature for the species's growth, heat will inhibit microbial growth. Pathogens normally will not grow in food that is held above 140°F (60°C). Organisms are killed as the temperature is increased further. The lethality depends on two physical factors: the temperature of the product and the length of time the product is held at that temperature. Increasing a product's temperature reduces the amount of time necessary to kill pathogenic cells. Vegetative microorganisms can be killed when a product is heated to temperatures of less than 212°F (100°C). If a low-acid food product is hermetically packaged, the product must be commercially sterilized to kill spores of *C. botulinum.* Temperatures used to kill bacterial spores are usually in excess of 250°F (121°C). A process authority must develop the specific time–temperature relation for commercial sterilization processes. For example, the FDA and USDA recognize organizations such as the National Food Processors Association, as well as competent personnel in the Cooperative Extension Service, as having expertise to develop these processes.

Pasteurization can be defined as the time–temperature relation that provides sufficient lethality to kill all vegetative pathogenic microorganisms and reduce the nonpathogenic microorganisms to an appropriate level in a specific food product. Commercial sterilization can be defined as the time–temperature relation that provides sufficient lethality to kill all pathogenic microorganisms (both vegetative cells and spores) in a specific food product. Commercial sterilization is not designed to totally sterilize a food. Commercially sterilized foods may contain thermophilic microbial spores that are not pathogenic.

Cold

A decrease in temperature below the optimal growth temperature for the species will reduce the growth rate of organisms. Most microorganisms that can affect food safety will not grow at temperatures below 40°F (4°C). However, food must be frozen to completely stop the growth of microorganisms. Most foods start to freeze at temperatures of 28°F (–2°C). Frozen food is held at temperatures of less than 0°F (–18°C) to inhibit undesirable chemical reactions. Freezing food may reduce the level of viable microorganisms in a food product. However, refrigeration or freezing rarely can be used as a control strategy to eliminate a foodborne disease problem.

Chemicals Used in the Food Processing Plant

Food processors use other chemicals at the processing plant to aid in the control of biological hazards.

The cleaning of food processing equipment can be classified as both a physical and chemical method that is used to control the growth of microorganisms. Cleaning physically removes residual foods that can be a source of nutrients and water for microorganisms. In addition, cleaning agents can physically kill microorganisms by disrupting their cells. If cleaning is not done periodically, the food processing equipment will serve as a source of contamination for future lots of food. Equipment must be cleaned prior to sanitizing.

Sanitization is a chemical process used to kill viable microorganisms that may be left on equipment after the cleaning process. A number of sanitizing compounds have been approved for use in food processing plants including chlorine, iodophores, and quaternary ammonium compounds. Sanitization must take place to reduce the chance of cross-contamination. It is not a substitute for proper cleaning of equipment or facilities.

Pesticides are used in processing plants as part of an insect and rodent control program. Insects and rodents harbor microorganisms, add filth to food, and can cross-contaminate food. To ensure the safe production of food, these chemicals must be stored and used in a manner that complies with federal regulations. The proper use of these chemicals is necessary to ensure the HACCP pest control prerequisite program (PRP) is effective.

Physical and Mechanical Control

In addition to controlling temperature and water activity (A_w), food processors can use other mechanical means to control microbial growth.

Packaging material provides a physical barrier that can be used to reduce microbial spoilage. Packaging material protects food products from microbial contamination during storage, transportation, distribution, and retail sale. Sources of contamination include handling by humans, environmental contamination, and cross-contamination by pests. Packaging systems have been developed to control the atmosphere of fresh-packed foods such as ready-to-eat vegetable salads. With this type of packaging, the oxygen and carbon dioxide levels are controlled at levels that inhibit microbial growth.

Radiation

Ionizing radiation is used to cold-pasteurize foods without changing the character of the raw food product. Ionizing radiation disrupts the genetic material of living cells and kills parasites, insects, and pathogenic bacteria. In addition, radiation reduces the level of nonpathogenic bacteria in food. Food irradiation does not make the food radioactive. In recent years, the FDA has approved the irradiation of a number of foods including spices, vegetables, poultry, and red meat.

Chemical Hazards

Food can become contaminated with chemical hazards during any stage of the manufacturing process. There are a number of different types of chemical hazards that can occur, of natural and human origin.

Some food components cause allergic responses in sensitive individuals. The primary causal agents in allergic reactions are proteins. Food proteins implicated in food allergy reactions come from a wide variety of sources including peanuts, tree nuts, eggs, milk, shellfish, wheat, and soy. Other chemicals implicated as allergens include some artificial colors and sulfites. The primary control measures for allergens are the proper labeling of food and the proper cleaning of food processing equipment after a product that contains a known allergen has been produced. It should be noted that when the Food Safety Modernization Act (FSMA) is fully implemented, allergens will have to be treated as a distinct hazard separate from chemical hazards.

Plant or animal tissue can be contaminated with chemical hazards by environmental contaminants or the improper use of agricultural chemicals such as pesticides, antibiotics, or hormones. A primary source of environmental contamination is contaminated water. For this reason, some HACCP PRPs include a water testing requirement.

Food also may be contaminated in the manufacturing plant. Most of these problems can be attributed to contamination with industrial chemicals, such as cleaning agents, sanitizers, pesticides, and lubricants. Control measures include ensuring the proper use and storage of these chemicals. Food additives should not cause a chemical hazard when added to foods at the proper levels and used within limits established by regulatory agencies.

Physical Hazards

Physical hazards can enter a food product during any stage of the production (farming), manufacturing, or distribution process. Mortimore and Wallace classify physical hazards as:

- Items that are sharp and could penetrate the skin or gastrointestinal tract
- Items that are hard and could cause damage to teeth
- Items that are capable of blocking the respiratory tract and could cause choking[2]

Many sources exist for physical hazards. These sources include inadvertent contamination on the farm (such as stones, insects, wood, and dirt), inadvertent contamination during processing (bone fragments, wood, glass, plastic, metal fragments, and so forth), and contamination during distribution because of packaging failure. In the United States, the FDA states that the presence of poisonous or deleterious substances in food makes the food adulterated. However, the FDA also recognizes that some substances may be unavoidable contaminants in food since they may be necessary for food production or may be an inherent component of the food. Therefore, tolerances have been set at levels that pose no inherent hazard to human health. Regulatory action can be taken only if the defect level is exceeded. However, it is illegal for a food processor to knowingly blend out a contaminant to reduce its level in a food product so it is below the regulatory limit.

It should be noted that when the FSMA is fully implemented, organizations in the food supply chain will have to address radiological hazards. Radiological hazards will have to be addressed as a separate hazard.

Control measures for physical hazards include using inspection and sorting procedures to remove contaminants and maintaining the proper preventive maintenance systems to ensure that food is not contaminated during production, packaging, storage, and delivery.

NEW FOODBORNE DISEASES AND NEW TECHNOLOGIES

Food safety and the control of pathogenic microorganisms is not a static science. In recent years, food microbiologists have identified a number of pathogenic microorganisms including *Campylobacter jejuni, Yersinia enterocolitica,* and *E. coli O157:H7.* In addition, food technologists currently are developing a number of processing strategies to control microbial contamination in food. Strategies under development include ohmic and inductive heating, microwaves and radio frequency waves, high-voltage arc discharge, pulsed electric fields, oscillating magnetic fields, pulsed light technology, ultrasound, high-pressure processing, and pulsed x-rays.

To effectively audit HACCP programs, professionals must be aware of new developments in food safety and control mechanisms.

NOTES

1. U.S. Food and Drug Administration (FDA), *Bad Bug Book: Handbook of Foodborne Pathogenic Microorganisms and Natural Toxins,* 2nd ed., 2012, http://www.fda.gov/Food/FoodborneIllnessContaminants/CausesOfIllnessBadBugBook/ucm2006773.htm.
2. Sara Mortimore and Carol Wallace, *HACCP: A Practical Approach,* 3rd ed. (New York: Springer, 2013): 87.

Chapter 17

Meat and Poultry

PROCESSING CATEGORIES FOR MEAT AND POULTRY

In 1996, the USDA/FSIS implemented the HACCP and pathogen reduction final rule. This rule requires all meat and poultry processors—with the exception of custom processors—to develop and implement working HACCP plans for their products. Companies also are required to develop written sanitation standard operating procedures (SSOPs) and generic E. coli testing programs for specific products. In addition, the USDA/FSIS mandates that companies satisfy salmonella testing standards set forth for various meat and poultry products.

In the final rule, the USDA/FSIS also specifies certain basic requirements that meat and poultry plants must include in their HACCP plans. For example, the plans must: (1) address the seven principles of HACCP; (2) identify the intended consumers of the product; and (3) create a flow diagram of the product process. In addition, any employee who develops, modifies, or reassesses a company's HACCP plans must successfully complete a course of instruction in the application of the seven HACCP principles. This training must include sections on the development of a HACCP plan for the specific product and record review.

When developing HACCP plans for meat and poultry products, companies should first group similar products together. Examining the approved labels of the products and categorizing products according to ingredients and processing parameters (such as raw, heat-treated, ready-to-eat, and so on) normally is the easiest way to do this. USDA/FSIS requires each HACCP plan to place meat and poultry products into one of nine processing categories:

1. Slaughter—all species
2. Raw product—ground
3. Raw product—not ground (for example, meat cuts, whole or cut-up birds)
4. Thermally processed—commercially sterile (such as canned soup)
5. Not heat treated—shelf stable (such as jerky)
6. Heat treated—shelf stable (such as edible fats)
7. Fully cooked—not shelf stable (such as ham)

8. Heat treated but not fully cooked—not shelf stable (such as char-marked beef patties)
9. Product with secondary inhibitors—not shelf stable (such as fermented sausage)

Plants may develop a single HACCP plan for one processing category to cover multiple products provided that the food safety hazards, critical control points (CCPs), critical limits, and processing procedures are similar. For example, a company may produce various types of hot dogs (all beef, beef and pork, and so on) under the same HACCP plan. However, a deviation in a CCP for one product may adversely affect other products under the same HACCP plan. If monitoring determines that the internal cooking temperature of the all-beef hot dogs was less than the required critical limit, other types of hot dogs produced might also be suspect.

Similarly, companies may develop a single HACCP plan to encompass more than one processing category. For example, a plant may slaughter and fabricate whole muscle cuts. Therefore, it might be more efficient to have one HACCP plan covering both processing categories ("slaughter—all species" and "raw product—not ground").

HACCP PLAN DEVELOPMENT

As discussed in Chapter 2, certain preliminary tasks must be completed before the HACCP principles are applied to specific products or processes. Once the HACCP team has been assembled, meat and poultry processing plants should determine the possible number of HACCP plans to develop by dividing their products into the nine processing categories. After the consumers and their intended use of the food have been identified, a detailed process flow diagram should be developed. All processing steps directly under the control of the company should be included, along with steps occurring before and after the processing stage. All ingredients and packaging material associated with the product from receiving to processing to distribution should be included in the flow diagram.

The receiving step should be detailed enough to distinguish between the receiving of dry ingredients, packaging materials, and perishable meat and non-meat items. In meat and poultry operations, a flow diagram should not just have a process step called "receiving of meat and/or poultry." It is more beneficial to break down this step into three categories: (1) receiving of refrigerated raw meat and/or poultry; (2) receiving of frozen raw meat and/or poultry; and (3) receiving of cooked meat and/or poultry (refrigerated or frozen). Potential hazards associated with the different processing categories during the receiving step are more efficiently addressed in this manner.

Once items are received, they often are stored for a period of time. Meat and poultry products frequently are tempered; that is, product temperature is adjusted from a storage temperature to a processing temperature prior to the start of production. Therefore, storage of all ingredients, along with packaging and any tempering steps, should be included in the flow diagram.

Movement of product from one processing area to another is also a critical detail in a flow diagram. An example of product movement would be opening boxed beef, portioning beef in one room, and then moving the portioned beef to another room for final packaging. A company may decide to address the potential hazards and CCPs due to movement of product in various ways. For example, if rooms in a plant are on different refrigeration units, a CCP may be implemented that measures either the room or product temperature for each room.

In slaughter operations, movement of carcasses into hot coolers or aging coolers and fabrication should be addressed along with the processing steps of edible by-products. The flow diagram for slaughter operations should specify where the process for slaughter ends if the operation also fabricates carcasses and includes grind operations. One approach may be to end the process after carcasses are chilled in the cooler for a specific period of time or removed for fabrication.

In grinding and sausage operations, the first (coarse) and second (fine) grindings should be steps in the flow diagram along with other equipment/processes such as patty machines, choppers, emulsifiers, and stuffers. Each of these steps can potentially increase the temperature of the meat and/or poultry products and may be the best places for CCPs, since further processing may increase the temperature of the final product to unacceptable levels.

As the final preliminary task in HACCP development, the flow diagram should be verified through observation of the processing steps in the plant environment. Additions or corrections to the process can be made at this time. By developing a very detailed and accurate process flow diagram, the company will be better equipped to apply the seven principles of HACCP, beginning with the hazard analysis and identification of potential CCPs.

Hazard Analysis

The USDA/FSIS requires that companies conduct a hazard analysis to determine potential food safety hazards that are reasonably likely to occur in the production process. This includes food safety hazards that can occur before, during, or after entry into the company's processing operations. Preventive measures for controlling the identified food safety hazards also must be determined. In reviewing a company's hazard analysis, the USDA/FSIS checks that the food safety hazards are identified, that preventive measures are listed, and that this information is in the hazard analysis section of each HACCP plan.

To determine the potential food safety hazards and possible preventive measures, the following three questions can be asked for each processing step in the flow diagram:

1. *What are the potential biological, chemical, and physical hazards?* In meat and poultry operations, biological hazards include pathogens such as *Salmonella spp., E. coli O157:H7, Listeria monocytogenes,* and *Campylobacter jejuni.* Chemical hazards include cleaners, sanitizers, lubricants, and products containing restricted ingredients such as nitrite. Buckshot, BBs, plastic, needles, metal fragments, and bone chips are physical hazards frequently associated with meat and poultry products.

A common tendency when listing potential hazards is to write "unclean equipment," "cross-contamination," "improper cooking temperature," or "excess sanitizer residue." These problems may cause biological or chemical hazards but are not the actual hazards. At a minimum, a company should list pathogens (biological hazards), foreign objects (physical hazards), and cleaners and sanitizers (chemical hazards) that could come in contact with products. This list should be updated as new hazards are identified.

2. *Are these potential hazards significant enough and likely enough to occur that they should be addressed in the HACCP plan?* Once the hazards are listed, the questions of whether the risk associated with a particular hazard is significant and whether that hazard is likely enough to occur that it should be included in the HACCP plan need to be answered. Hazards that have a low incidence rate, have a low likelihood of occurrence, or are controlled by an effective preventive measure in a later processing step are not normally associated with HACCP plans. In addition, if the hazard is controlled by a prerequisite program (PRP) such as an SSOP or a company standard operating procedure, it usually will not be included in the HACCP plan.

3. *What information/method has been used to justify the decision concerning the significance of a particular hazard?* In slaughter operations, antibiotics and hormones are chemical hazards associated with the processing step of receiving live animals. A company may justify that this is a low-incidence hazard because of USDA/FSIS or state monitoring programs. Another justification may be that the company purchases animals from a limited number of suppliers, requiring a form signed by the producers to assure appropriate drug use and withdrawal times.

In processed meat and poultry operations, receiving of raw refrigerated products is a safety concern, especially if the company processes ground products. Even if the company is buying meat and poultry from another plant operating under HACCP, temperature abuse could occur during transportation. Some operations may believe that the likely occurrence of temperature abuse is low and have monitoring devices to record temperatures while product is in transport. Others may maintain that monitoring temperatures during receiving is a preventive measure for ensuring temperature control to prevent growth of pathogens to unacceptable levels.

In ready-to-eat meat and poultry operations, a common justification during processing steps such as grinding and stuffing is that thermal processing occurs at a later step in the process. A potential biological hazard during packaging of ready-to-eat meat and poultry products is *Listeria monocytogenes* contamination from equipment. An effective SSOP program and plant program addressing personal hygiene (listed as preventive measures) or a cooking step after packaging and before consumption may be used as reasons for not addressing this biological hazard in the HACCP plan, provided the label clearly says "cook before consuming product."

In meat and poultry operations, SSOPs may be used to control chemical hazards from excess residue of cleaners and sanitizers or biological hazards from unclean equipment. Effective company programs for sanitation, personal hygiene, receiving and storage of perishable items, and pest and rodent control may be used to address specific biological, chemical, and physical hazards. These PRPs then reduce the number of possible CCPs in a HACCP plan. However, companies must ensure that these programs are effective and are being maintained on a routine basis.

Critical Control Points

Once the hazard analysis has been conducted, companies must determine the CCPs for each HACCP plan. According to the USDA/FSIS, all meat or poultry processes will encounter one or more food safety hazards that are deemed significant and that must be addressed in the HACCP plan. Therefore, a HACCP plan must contain at least one CCP.

In the HACCP plan, any processing step that is a significant source of hazards should be considered as a possible CCP. A CCP is any point or step where a food safety hazard can be prevented, eliminated, or reduced to acceptable levels.[1] For instance, in fabrication and grinding operations, controlling the temperature of raw product during processing only prevents pathogens from increasing to unacceptable levels. In contrast, fermentation or acidification reduces, but does not eliminate, pathogens. A thermal processing step, on the other hand, either eliminates pathogens or reduces pathogens to an acceptable level.

In slaughter operations, the USDA/FSIS requires plants to have a CCP addressing "zero tolerance" of fecal contamination. The primary mode of removing contamination must first be knife trimming. Acid sprays, hot water sprays, steam vacuuming, and steam pasteurization are often used as additional methods for reducing pathogens that may be associated with fecal contamination. Plants utilizing these processes may develop a CCP that monitors the temperature of the steam or water or the pH of the acid spray.

Critical Limits

Once a CCP has been identified, a critical limit should be set to determine whether a CCP is in or out of control. Exceeding critical limits may indicate that a direct health hazard exists, a direct health hazard could develop, or product was produced in an environment where direct health hazards may have been present. In meat and poultry operations, the USDA/FSIS has many specified time and/or temperature requirements for products such as precooked ground beef, roast beef, and ham. In addition, pH and moisture-to-protein ratios are defined by USDA/FSIS for specific products such as fermented sausages, jerky, and snack sticks. At a minimum, these requirements must be used by companies to determine critical limits in their HACCP plans. Published and reviewed research papers also provide information on effective processing parameters for controlling pathogens in meat and poultry products.

When setting critical limits, companies need to set values that are "greater than or equal to" or "less than or equal to" and avoid ranges. For instance, it is better to specify a final internal temperature for hot dogs of ≥165°F (≥74°C) than 165°F (74°C). If this were not done and the temperature measured was 165°F (75°C), then corrective action in regulatory terms would be required even though the product was considered to be safe. In addition, a company may decide to have a higher internal temperature than required by the USDA/FSIS in order to extend shelf life. However, a company may decide not to use the higher temperature as a HACCP critical limit because the USDA/FSIS lower temperature ensures a safe product. Also, if the higher temperature is set as a critical limit in the HACCP plan, then this temperature will become the USDA/FSIS regulatory limit.

Monitoring

The USDA/FSIS requires that meat and poultry establishments define monitoring procedures for each CCP and specify the frequency with which the procedures will be performed. A series of five questions can assist in developing detailed monitoring procedures. Following are the questions and an example of an appropriate response for each:

1. *What will be measured?* Internal temperature of ham.
2. *Where will the critical limit be measured*? In the center of the largest ham in the smokehouse.
3. *How will the critical limit be measured?* Using a calibrated smokehouse internal thermometer with a recording chart.
4. *Who will monitor the critical limit?* Smokehouse operator (use job titles, not names of specific individuals).
5. *How often will the critical limit be measured?* Every batch.

Continuous monitoring, such as using a thermometer with a recording chart, is preferred over interval monitoring. However, continuous monitoring is not always possible. For instance, a ground beef or fresh sausage processor may monitor a product's temperature after the first grind. Depending on the company's operations, frequencies to check internal product temperature may be every half hour, every batch, or four times a day. Once the interval is set, it becomes a regulatory requirement. Therefore, companies must be capable of monitoring at the set frequencies. In addition, a company must realize that if the specified critical limit is not met, all products since the last check will be in question.

Corrective Action

If monitoring indicates a deviation, corrective actions are required to ensure product safety. The USDA/FSIS requires that the following four action points be addressed for every corrective action:

1. The cause of the deviation is identified and eliminated.
2. The CCP will be under control after the corrective action is taken.
3. Measures to prevent reoccurrence have been established.
4. No product that is injurious to health or otherwise adulterated as a result of the deviation enters commerce.

One simple approach to meeting regulatory HACCP requirements for corrective actions is to state, "The plant will adhere to regulation 417.3." Meat and poultry establishments may then have a separate record sheet with the aforementioned four corrective action points and use this sheet for documenting all corrective actions.

Other processors may decide that it is better to have a formalized approach for correcting deviations because they want their employees to follow certain guidelines when a deviation occurs. Another series of questions similar to the USDA/FSIS's questions may be used; these questions and some typical responses follow:

1. *What are some possible causes of deviations?* Equipment failure, human error.
2. *How could the process be corrected?* Cook for a longer period of time, place product on hold for further analysis.
3. *How will the product be handled/disposed of?* Send product to rendering, rework product, hold and release after test results are received.
4. *What measures could be implemented to prevent reoccurrence?* Develop a detailed maintenance program, retrain employee.
5. *Who is responsible for implementing the corrective action?* Plant manager, supervisor.

Verification

The USDA/FSIS requires plants to list verification procedures and the frequency with which they will be performed. Examples of verification activities include calibration of monitoring instruments, observation of monitoring procedures, review of records, and microbiological testing. A slaughter operation may use data from its E. coli testing program as a verification procedure for control of fecal contamination. A jerky operation may routinely analyze protein and moisture content to ensure that the green weight loss used during monitoring achieves a specified moisture-to-protein ratio.

Calibration of equipment is an important verification procedure in all HACCP plans. Setting the interval for calibration is also important. If a company calibrates on a daily basis and determines the next morning that the thermometer was reading higher-than-normal values, then all product produced the day before could be in question.

Record Keeping

Record keeping is the essential component of HACCP. The USDA/FSIS requires that records contain "the actual values and observations obtained during monitoring."[2] This includes actual times, temperatures, or other quantifiable values; the calibration of instruments; corrective actions; and verification results. All records are to be signed and dated, and a product code or slaughter production lot should be defined. In addition, meat and poultry establishments must review all records prior to shipment of product.

Records for slaughter establishments and companies producing refrigerated meat and poultry products must be retained for at least one year, whereas records for frozen, preserved, or shelf-stable items require two years of retention. Processors may store records off-site after six months as long as they can be retrieved within 24 hours' notice by the USDA/FSIS.

Processors have the flexibility to develop their own record forms. Although a plant may have four different HACCP plans, one form may be used for corrective actions and another form for calibration of instruments. If room and product temperatures are CCPs for various HACCP plans, then a temperature recording log could be developed. Some companies have automated chart recorders for items such as internal temperatures of smokehouse products, room temperatures, temperatures of steam pasteurization units, and acid spray cabinets. The information obtained from the chart recorder is an acceptable record for USDA/FSIS requirements as long as the plant signs and dates the record and identifies the production code.

When recording data, the record should have a place for a person's signature or initials. Initials are acceptable, but a company should have a file identifying the person's initials. A signature stamp is not acceptable. The time, date, and actual value and/or observation should be recorded by the person monitoring the specified CCP. It is also helpful to have a simple monitoring procedure and the critical limits for various products on the record form.

Data should be entered at the time they are collected. Data should be recorded with a pen; if an entry error occurs, a single line should be drawn through it and the mistake initialed. Since the frequency required for monitoring may be every two hours, it is also important to record any downtimes during the production shift. There should be a place for the record reviewer's signature since preshipment review of records is required by the USDA/FSIS.

Reassessment of the HACCP Plan

The USDA/FSIS requires that meat and poultry companies reassess their HACCP plan on an annual basis.[3] The USDA/FSIS may announce that companies must also conduct a reassessment of their HACCP plans in the event of a change in regulatory policy or as a result of lessons learned from a foodborne disease outbreak or other regulatory action. In addition, a company should reassess its HACCP plan as a result of significant changes in its manufacturing processes. This can include changes in raw materials, process flows, equipment, product formulation, and product or packaging specifications. The USDA/FSIS outlines the process of reassessment in 9 CFR 417.4. The objective of a reassessment is to ensure the HACCP plan is functioning as originally intended. As a result, the reassessment should be an in-depth evaluation of the HACCP plan and must be properly documented.

NOTES

1. National Advisory Committee on Microbiological Criteria for Foods (NACMCF), *Hazard Analysis and Critical Control Point Principles and Application Guidelines* (Washington, DC: U.S. Food and Drug Administration, August 14, 1997).
2. *Code of Federal Regulations,* Hazard Analysis and Critical Control Point (HACCP) Plans—Fish and Fishery Products, title 21, sec. 123.6 (2006).
3. *Code of Federal Regulations,* Validation, Verification, and Reassessment, title 9, sec. 417.4.

Chapter 18

Seafood

Seafood includes all products located in the deep ocean as well as those found in freshwater rivers, streams, lakes, and estuaries. It sometimes includes products such as frog legs and alligator, depending on the jurisdiction of the agency involved. Seafood encompasses a large variety of species and product forms, making generalizations difficult. Many seafood products are cooked, some products are consumed raw, and, as in the case of oysters and clams, the body of the animal may even be consumed as a whole. However, even with all the variations, seafood safety issues fall into very easily defined and narrowly focused areas. The products of safety concern include raw molluscan shellfish, species subject to the formation of histamine, cooked ready-to-eat products, and vacuum-packaged smoked fish.

Seafood also is unique in that it reacts in different ways than other meat products such as meat and poultry. First, since seafood is extremely sensitive to abuse, its quality and safety are easily compromised if it is handled improperly. Second, microbes associated with seafood can grow at a wide variety of temperatures, including refrigerated temperatures. Third, seafood struggle as they are caught, as often they are wild harvested. This struggling uses up the glycogen in their muscles with no corresponding drop in muscle pH. As a result, microbial growth is not discouraged. This makes it critical to control postharvest product temperatures. Finally, the chemical composition of the fat in seafood is high in phospholipids, which break down into trimethylamine. This is the source of the "fishy" odor often associated with seafood. These fats also are highly unsaturated and are easily oxidized, resulting in additional off flavors. In general, seafood tends to deteriorate organoleptically before it becomes unsafe to eat.

SOURCES OF HAZARDS SPECIFIC TO SEAFOOD

The sources of hazards in seafood are varied and fall into three traditional categories: biological, chemical, and physical.

Biological Hazards—Bacteria

Campylobacter jejuni. Transmitted through contaminated raw clams, mussels and oysters, and contaminated water. The hazard can be controlled by thoroughly cooking seafood and by proper sanitation.

Clostridium botulinum. Found in the intestinal tracts of fish and the gills and viscera of crabs and other shellfish. Type E is most common in fish and fishery products, as it grows at temperatures as low as 38°F (3°C) and produces little evidence of spoilage. Proper thermal processes in canned seafoods, heavy salting or drying, or acidification are effective control measures.

Escherichia coli. Naturally found in the intestinal tracts of all animals, it is transferred to seafood through sewage or pollution, or by contamination after harvest through human handling. Control measures include heating sufficiently to kill the organism, holding chilled foods below 40°F (4°C), and preventing cross-contamination.

Listeria monocytogenes. Usually found through cross-contamination of cooked ready-to-eat foods by raw foods. Widespread in nature, it is controlled most effectively by good sanitation practices and thorough cooking of seafood products.

Salmonella spp. Found in the intestinal tracts of animals but not in fish. It typically is transferred to seafood through sewage or pollution, or by contamination after harvest. The hazard can be prevented through sufficient heating, proper holding of chilled products, and prevention of cross-contamination.

Shigella spp. Naturally found in the intestinal tract of humans, it is transferred to seafood through sewage or pollution, or by contamination after harvest. The hazard can be controlled by a proper water supply and preventing ill employees from coming in contact with the food.

Staphylococcus aureus. Commonly found in the nose, throat, and hair of humans, this bacterium finds its way into food through cross-contamination and through improper human contact with the food. Minimizing time/temperature abuse of seafood and proper food handler hygiene are steps that can be taken to prevent the hazard.

Vibrio cholerae. Found in estuaries and brackish water, it tends to be more prevalent in the warmer months. The hazard can be prevented by thoroughly cooking seafood and minimizing cross-contamination.

Vibrio parahaemolyticus. Naturally occurring in estuaries and coastal areas and also more prevalent in the warmer months. The hazard can be controlled by thoroughly cooking seafood and preventing cross-contamination.

Vibrio vulnificus. Requires salt for survival and primarily is found in the Gulf of Mexico. The hazard is more prevalent in the warmer months and can be prevented by thorough cooking of shellfish and by preventing cross-contamination.

Yersinia enterocolitica. Naturally found in soil, water, and domesticated and wild animals. Sufficient heating, proper chilled holding, and prevention of cross-contamination are effective methods of preventing the hazard.

Biological Hazards—Viruses

Hepatitis A virus. Survives better at low temperatures and is killed at high temperatures. Raw and steamed clams, oysters, and mussels have been implicated in contamination with this virus. The hazard can be prevented by thoroughly cooking the product and preventing cross-contamination. But since this virus seems to be more resistant to heat than other viruses, steaming mollusks only until the shells open typically will not inactivate the virus.

Norovirus. Considered a major cause of nonbacterial intestinal illness, it is associated with eating clams, oysters, and cockles. The hazard can be prevented by thorough cooking of seafood and preventing cross-contamination.

Biological Hazards—Parasites

Anisakis simplex. A parasitic nematode, it is found in undercooked or raw (sushi, sashimi) fish. Parasites are a hazard only if the fish is to be consumed raw. Specific freezing processes will kill the parasite.

Pseudoterranova decipiens. Also called codworm, it is a nematode found in raw or undercooked fish. Control is the same as for *Anisakis simplex.*

Diphyllobothrium latum. A tapeworm that infects primarily freshwater fish and is also found in salmon. Found in raw or uncooked fish, the control is as discussed previously.

Chemical Hazards—Marine Biotoxins

Amnesic shellfish poisoning (ASP), diarrhetic shellfish poisoning (DSP), neurotoxic shellfish poisoning (NSP), and paralytic shellfish poisoning (PSP) have been associated in different waters with molluscan shellfish and at times small species such as herring, anchovies, and crustaceans such as crabs. The control method is typically closing and monitoring of the harvest waters, as these hazards cannot be removed from seafood product once present.

> *Ciguatera fish poisoning (CFP).* Found in some species of tropical and subtropical fish such as red snapper. Again, the control mechanism is careful monitoring of the catch, utilizing local knowledge of the harvest area or closing the waters.
>
> *Scombroid toxin (histamine).* The hazard is caused by eating certain species of fin fish—including tuna, mahi-mahi, bluefish, sardines, amberjack, and mackerel—that have gone through spoilage by certain types of bacteria. The toxin is not eliminated by cooking or canning, and typically is controlled through proper chilling from harvest to process, utilizing a detailed knowledge of the temperature history.
>
> *Tetrodotoxin.* A potent toxin associated with puffer fish. These fish cannot be imported into the United States without authorization. In other countries, such as Japan, certification of chefs preparing the fish is required.

Other chemical contaminants include aquaculture drugs, chemical contaminants (pesticides, herbicides, and so on), and food additives such as bisulfites in shrimp. These contaminants are controlled either by proper withdrawal times, as in aquaculture drugs, periodic testing, and approved suppliers, as in pesticides, or monitoring and testing, as in bisulfites.

Physical Hazards

Metal fragments, glass, wood, plastic, and so forth find their way into seafood products from time to time, especially breaded fish portions, fish sticks, and breaded shrimp. Metal is the only physical hazard that is easily detected—through the use of metal detectors inline or after packaging.

Emerging Hazards

The recent Food Safety Modernization Act (FSMA) and the subsequent fourth edition of the *Fish and Fishery Products Hazards and Controls Guide* have expanded hazards to include allergens and radioactivity. All seafood products must include information on the label that allergens are present, and this activity must be controlled by the firm's HACCP plan. Radioactivity, especially in seafood as a result of the 2011 Fukushima accident, is a sensitive issue in the seafood industry. Although no evidence has been found that radiation toxicity has reached actionable levels, firms must continue to consider this issue in their hazard analysis.

HACCP REGULATIONS IN THE UNITED STATES

The FDA is the primary federal agency responsible for the safety of seafood in the United States. Through the Food, Drug, and Cosmetic Act, the FDA has the responsibility to ensure that all domestic and imported seafood products are safe, wholesome, and properly labeled. In December 1997, the FDA published its final Seafood HACCP Rule, 21 CFR part 123, defining the HACCP and sanitation requirements of seafood facilities.

The FDA developed a highly prescriptive HACCP program for the seafood industry. The seafood HACCP regulations require prerequisite programs (PRPs). These programs must comply with good manufacturing practices (GMPs) (21 CFR 110), sterilization standard operating procedures (SSOPs), monitoring and record keeping for eight areas of sanitation, and a documented HACCP plan. The FDA published a list of potential hazards and control limits in the *Fish and Fishery Products Hazards and Controls Guide.*

The FDA works with state regulatory agencies to ensure implementation of federal regulations. For example, in California the FDA and the State of California Department of Health Services Food and Drug Branch (CDHS-FDB) share jurisdiction and regulatory oversight of all seafood manufacturers and distributors. Federal and state agencies collaborate to provide joint food safety training and regulatory enforcement. This partnership allows the regulatory agencies to develop a shared work plan and common enforcement strategy. The state regulatory agencies may exert a more restrictive enforcement action, such as embargoing or recalling potentially unsafe food. An embargo may take place if appropriate and may be mandated in situations where illnesses have been caused by the lack of HACCP plans, lack of GMPs, or lack of SSOPs.

FDA regulations require that the individual developing the plan be knowledgeable and trained in the principles of HACCP. This can be done by completing a HACCP course with a curriculum that is recognized as adequate by the FDA or through HACCP-relevant job experience.

During the initial implementation of the HACCP program in the seafood industry, the FDA identified a number of problems, including the following, during inspections:

- The corrective action process did not meet the regulatory requirements, usually because: (1) it did not adequately prevent adulterated product from entering the market; or (2) it did not correct the root cause of the deviation.
- The HACCP plan did not adequately distinguish who was responsible for the control of specific hazards. For example, some hazards were controlled by the companies that harvested seafood, while other hazards were controlled by the companies responsible for further processing of the seafood.
- The HACCP plan did not list specific control parameters—for example, the time and temperature parameters that must be met to prevent a food safety hazard.
- Weak linkages existed between the critical limits and monitoring procedures. For example, measuring and recording a cooler temperature is not an adequate method for monitoring product temperature unless a validation study has been conducted under a worst-case scenario.
- Critical limits were set without being scientifically supported. One source of critical limits is the FDA's *Fish and Fishery Products Hazards and Controls Guide.*

The FDA developed a training program titled "Seafood HACCP Encore Course" to improve the HACCP implementation process in the seafood industry and further define the HACCP seafood regulations.

The National Marine Fisheries Service (NMFS) operates the National Oceanic and Atmospheric Administration Seafood Inspection Program (NOAA SIP). Through the Agricultural Marketing Act, NMFS operates this voluntary, fee-for-service inspection program. Typical services include sanitation evaluation of facilities, product inspection and certification, and reduced inspection programs such as the HACCP Quality Management Program (HACCP QMP), where a firm's adherence to the principles of ISO 9001 and ISO 22000 are audited using internationally recognized auditing practices.

The U.S. Department of Commerce believes that HACCP-based services enhance the safety, wholesomeness, economic integrity, and quality of seafood available to consumers, as well as improve seafood industry quality assurance and regulatory oversight.

Other federal agencies with jurisdiction over seafood include the U.S. Customs Service, the Environmental Protection Agency (EPA), and the USDA. Each has some regulations associated with the labeling of seafood, the tolerances of pesticides, and so forth. However, these agencies do not have specific programs designed to assure safe, wholesome, and properly labeled seafood.

APPLIED HACCP VERSUS QUALITY

By far the biggest controversy in seafood HACCP has been the debate over the use of quality programs to aid or oversee HACCP systems in facilities. However, many companies consider the use of HACCP systems in harmony with quality systems to be the best approach with regard to seafood, as seafood is generally a safe product to eat. This concept is outlined in the Codex Alimentarius Commission HACCP Guidelines, where critical control points (CCPs) and critical limits are associated with HACCP and food safety, while defective action points (DAPs) and control limits are associated with quality and regulatory defects in the product. The development of PRPs, management commitment, and employee education and training in HACCP principles are some of the factors that contribute to the success of a HACCP program.

Prerequisite Programs

For HACCP to function effectively, it must be performed in conjunction with PRPs that provide a foundation of compliance for the company. These PRPs include GMPs and acceptable SSOPs. The GMPs (21 CFR 110) are broad in focus and define measures of general hygiene, along with measures the company should take to prevent food adulteration due to unsanitary conditions. SSOPs are procedures used by the company to define how it will accomplish adherence to the GMPs. As required by 21 CFR part 123, the SSOPs should be written and monitored. Some companies integrate them into the company's quality program. GMPs and SSOPs affect the overall processing environment and thus do not lend themselves readily to a specific CCP. Therefore, they work best in a procedural form with directions for the implementation, monitoring, and corrective action of plant and food hygiene concerns. In fact, well-defined and well-monitored SSOPs can reduce the identification and number of CCPs by controlling, for example, cross-contamination and chemical contamination. Strong SSOPs make the HACCP system more effective by allowing the system to truly concentrate on the food safety aspects of the manufacturing process rather than on the processing plant environment. By far, the most common reason for the failure of a HACCP system in a seafood facility is related to sanitation. The most common sanitation failure is due to employee practice. These points should be considered when auditing any seafood HACCP system.

Management Commitment

For any quality or safety system, including a HACCP plan, to work, the support of top company officials is essential. Without this support, HACCP will not become a company priority or be effectively implemented. In fact, this is noted as a common root cause for the failure of many HACCP systems in seafood facilities.

HACCP Training

Education and training are important elements in developing and implementing a HACCP program. Employees who will be responsible for the HACCP program must be adequately trained in its principles. It is also required that certain elements of the HACCP plan and its implementation be performed by properly trained personnel. Skipping this step violates the HACCP seafood regulation and will lead the company down a very short path to a failed audit.

HACCP AND ECONOMIC INTEGRITY

Along with product quality, economic integrity is the key to successful marketing of seafood. Quality and integrity are what buyers and consumers want. Consumers often have a limited knowledge and understanding of fish and seafood products. Consequently, their greatest concern and challenge is to purchase products that are wholesome and that have acceptable sensory appeal (such as taste, odor, texture, flavor, and appearance).

Economic integrity means that the product adheres to codes, standards, or specifications set by the producer, buyer, or regulators. In other words, the product is what it appears or is supposed to be. Certain economic integrity issues, such as species and product identity, country of origin, and short weight (net weight) are regulated by local, state, or federal agencies. However, with dwindling resources, current enforcement strategies tend to place a low priority on these issues relative to food safety. Lapses in enforcement create a risk of consumers receiving mislabeled, overbreaded, overglazed, short weight, or possibly unsafe products. Such lapses often are noted by consumer groups, the media, and the seafood industry. Companies have a legal and moral obligation to provide products of high economic integrity to the buyer and consumer.

FOOD SAFETY MODERNIZATION ACT

The FSMA includes provisions on the use of third-party audits by the FDA in the regulation of foods imported into the United States as well as those produced domestically. The use of third parties for inspections of foreign seafood processing facilities will have a profound effect on the industry, as the majority of seafood consumed in the United States is imported. These third-party programs are designed to speed the entry of acceptable foods into the country and at the same time increase the inspection level.

Chapter 19
Dairy

SAFETY REGULATIONS IN THE DAIRY INDUSTRY

Dairy products have been an important element in human food since animals were first domesticated.[1] Originally "manufactured" within the home or on the farm, dairy products are now manufactured in small- or large-scale plants where the main goal is the production of safe, high-quality products. The food safety process for fluid milk is monitored by state and federal regulatory agencies. With less than 5% of reported foodborne diseases originating from contaminated dairy products, the U.S. dairy industry has an excellent safety record—especially when the volume of dairy products manufactured is considered.[2] This record has been achieved through the efforts and actions of the National Conference on Interstate Milk Shipments (NCIMS), which has allowed the development of very prescriptive regulations for the production of Grade A milk products. The NCIMS has started an initiative to implement a voluntary HACCP program. This increasing effort to make HACCP a standard practice within the dairy industry is due to anticipation that HACCP will eventually be a mandated food safety system for the industry.[3]

Pasteurized fluid milk has a very short shelf life. Federal regulations require that all milk receive a heat-treatment step (pasteurization) designed to kill all vegetative pathogenic microorganisms. If a fluid milk product is packaged under anaerobic conditions, the potential for *C. botulinum* growth must be controlled either through a sterilization step that renders the product commercially sterile or by reducing the A_w to less than 0.85.

It is imperative that dairy producers have strict controls over prerequisite programs (PRPs) to ensure both product safety and product quality. For example, many dairy products, such as yogurt, buttermilk, and cheese, are produced using microbial cultures. The presence of therapeutic drugs may inhibit the growth of these starter cultures. Therefore, antibiotic residues must be carefully controlled for product quality reasons as well as for product safety reasons.

The process for ensuring safety is outlined here. Milk usually is received via tanker trucks and unloaded. Each tanker is checked for proper temperature and the presence of therapeutic drug residues. The temperature of the milk should be less than 45°F (7°C), and it should be held for no more than 72 hours before processing. The milk is then filtered and stored in tanks or silos. When needed, the milk is separated; the raw cream is stored or shipped elsewhere for other uses, and the remaining milk is pasteurized. The raw cream temperature should also be less than 45°F (7°C), and it should be held for no more than 72 hours. The cream is added back to the raw milk prior to homogenization and pasteurization to adjust the fat content of the milk. Vitamins may be added at the proper concentrations either before or after pasteurization. The pasteurized milk goes into a storage tank from which it is packaged, refrigerated, and placed into distribution.

TYPES OF HAZARDS

Biological, chemical, and physical contamination of dairy products is a major industry concern. Biological hazards include *Brucella, C. botulinum, Listeria monocytogenes, Salmonella, E. coli O157:H7, Yersinia enterocolitica, Campylobacter jejuni, Staphylococcus aureus,* natural toxins, and parasites. Chemical hazards include food additives, allergens, and unintentionally added chemicals. Potential physical hazards include metal, glass, insects, wood, plastic, and personal effects. Appendix A contains lists of recognized food hazards.

Formulated dairy products such as cheese, ice cream, and yogurt may be prone to potential hazards. This can happen due to the following reasons: (1) formulated dairy products may require extensive handling in vats, batch tanks, fruit feeders, and other equipment that potentially expose the product to the environment; (2) ingredients are often added to the milk base after the milk has been pasteurized; and (3) the final, formulated dairy product mix usually does not receive a terminal heat treatment step.

CONTROLLING RISKS THROUGH PREREQUISITE PROGRAMS

As in other food processing areas, the establishment of effective PRPs will simplify any HACCP plan for the dairy industry. These guidelines assist in the day-to-day operation of a dairy plant and help ensure the manufacture of safe dairy products. PRPs may be used to reduce the likelihood of a potential hazard. If the potential hazard is reduced to the extent that it is *not* likely to occur, then a critical control point (CCP) is not necessary to control that specific potential hazard. However, the company needs to ensure that the PRP is properly designed, operated, and maintained. This is done through inspecting, checking, and documenting the program. Questions that can be asked to ensure the PRP is operating properly include:

- Who is performing the PRP?
- How were these operators trained?
- What are the specific processes used for the program?
- Where is the program being applied?
- When is the program being conducted?
- How can it be confirmed that the program is being carried out in an effective manner?

Areas covered by prerequisites may include supplier control programs; receiving/storage programs; premises, equipment performance, and maintenance programs; cleaning and sanitization of equipment and facilities; recall programs; allergen control programs; and personnel training programs. A brief overview of these areas as they pertain to the dairy industry follows.

Supplier Control Program

The dairy processor needs to ensure that dairy products are safe and wholesome. Milk must come from approved sources. During the formulation of a dairy product, hazards are most likely to be introduced while ingredients are being added. These ingredients may be added to a mix that contains pasteurized milk; however, the final mix may not receive a terminal pasteurization step. Therefore, these ingredients have the potential to be carriers of hazards. To reduce the potential for a food safety problem, plants may choose to purchase only from suppliers that have implemented a HACCP program.

Some companies develop control programs in partnership with suppliers to ensure understanding of and compliance with specifications. Supplier specifications and guarantees should be verified periodically. For example, periodic analyses should be run on incoming ingredients to ensure they meet specifications. Observations of the production area personnel using these ingredients should be documented.

Receiving/Storage Program

Dairy plants should have steps in place to prevent the occurrence of any product contamination. A program should be in place for storing finished product, as well as for inspecting trucks used for the shipping and receiving of products. The Code of Federal Regulations (21 CFR 110) requires inspection of incoming ingredients and packaging material.

Premises, Equipment Performance, and Maintenance Program

Dairy processing plants need to be designed in a manner that prevents any cross-contamination of product from occurring. Issues to be addressed include proper ventilation, appropriate lighting, and adequate and accessible hand-washing stations.

Plants need a written preventive maintenance program. One critical aspect is ensuring that equipment used for pasteurization or sterilization steps is operating properly and is controlled to ensure the manufacture of safe product.

Cleaning and Sanitation Program

The plant must have a pest control program and master cleaning and sanitation program in place. Sanitation should not have any negative impact on the safety of the product being manufactured. A well-designed and operated program should reduce the likelihood of the occurrence of environmental, chemical, and biological contaminants.

Recall Program

A recall program can be divided into three parts: traceability, recall system, and recall initiation. A mock recall should be performed annually, and a product coding system should be in place for ease of traceability.

Allergen Control Program

The plant must identify any ingredients that may contain an allergen. In addition, the plant must identify any products that do not contain allergens, especially non-dairy-containing products such as fruit juices or water. A written allergen-handling program should detail the control of allergens in the plant. The plant needs to work in conjunction with suppliers to determine which ingredients contain potential allergens. Allergen-containing products should be scheduled to run last on the production line to avoid cross-contamination, or sufficient time should be allowed for an allergen cleanup prior to the manufacture of a non-allergen-containing product. Some plants may treat allergens as chemical hazards in their HACCP plan.

Personnel Training Program

Documented training should occur on product handling and personal hygiene for all employees. Employee and visitor access should be controlled to prevent cross-contamination. Employees should be trained and encouraged to inform management of problems that may impact product safety.

Chapter 2 provides more detailed information about these and other prerequisite areas. As the food safety principles at the core of HACCP are applied throughout the dairy industry, the safety of an already safe product will only continue to increase.

NOTES

1. Marianne Smukowski and Norman Brisk, "Dairy Products Industry," in *Encyclopaedia of Occupational Health and Safety*, Vol. 3, 4th ed. (Geneva: International Labor Office, 1997).
2. Eric C. Johnson, John H. Nelson, and Mark Johnson, "Microbiological Safety of Cheese Made from Heat-Treated Milk," *Journal of Food Protection* 53 (1990): 441–52, 519–40, 610–23.
3. Brian W. Gould, Marianne Smukowski, and J. Russell Bishop, "HACCP and the Dairy Industry: An Overview of HACCP Costs and Benefits," in L. J. Unnivehr, ed., *The Economics of HACCP* (St. Paul, MN: Eagan Press, 2000).

Chapter 20

Fresh Fruits and Vegetables

DEFINING GAPS AND GMPS

Fresh fruits and vegetables are highly perishable and, in general, their quality cannot be significantly improved after harvest. Thus, managers of fresh produce operations should focus on quality maintenance through implementation of appropriate production and handling practices. Since food safety hazards associated with fresh produce are difficult to correct through remedial action, managers should focus their attention primarily on the prevention of hazard occurrence. A case study of the application of HACCP principles to a fresh market tomato handling operation appears later in this chapter.

The current body of food safety literature for fresh produce makes frequent reference to the proper implementation of good agricultural practices (GAPs). However, a close examination of the literature reveals that "GAP" is a nebulous term that refers as much to the use of common sense as it does to science-based management protocols. No single document serves as an adequate GAP reference for all the procedures involved in the production and handling of fresh fruits and vegetables. The FDA provides a guidance document that addresses GAP implementation for reduction of microbial contamination.[1] However, the information presented is general, and identifying appropriate, effective management practices remains the responsibility of managers.

In contrast, good manufacturing practices (GMPs) are much more clearly defined in the Code of Federal Regulations (CFR).[2] Guidelines are provided for personnel, buildings and facilities, equipment and utensils, and production processes. GMPs as they relate to prerequisite programs (PRPs) in the food industry were discussed in detail in Chapter 2.

HAZARDS ASSOCIATED WITH HANDLING FRESH PRODUCE

As in other areas of the food industry, potential hazards in the processing of produce can be divided into three categories: biological, chemical, and physical.

Biological

Microbial contamination (with human pathogens) has surpassed pesticide residues as the primary consumer food safety concern. Fresh fruits and vegetables differ from processed foods in that the handling of fresh produce does not entail a "kill step," such as sterilization, that renders a product commercially sterile. The consequence is that prevention of microbial hazard occurrence, as opposed to remedial treatments to eliminate hazards, is crucial during handling.[3]

Chemical

Historically, chemical (pesticide) residues represented the most significant food safety concern in consumer surveys. Agrichemical use is subject to EPA regulation with oversight (testing) by the FDA. This includes preharvest application of pesticides (fungicides, insecticides, and herbicides) as well as postharvest application of waxes, coatings, fungicides, or biocontrol agents used to prevent decay. Proper use of any agrichemical is a GAP and is not considered a critical control point (CCP) in a HACCP program. Failure to meet residue tolerance will result in outright rejection of the product by regulatory authorities. In the fresh-cut industry, chemical use may include sanitizers such as chlorine compounds and antioxidants. Use of sanitizers generally is considered a CCP.

Physical

Physical hazards generally are not a concern for whole fruits and vegetables, where the most common types of foreign materials in a package are soil, leaves, or a piece of a broken container—all of which are precluded by GAPs. The detection of physical hazards is, however, important in the fresh-cut, minimally processed industry, where pieces of cutting or packaging equipment can inadvertently be left in or comingled with the product. Closing machines that employ wire staples or similar materials can cause problems. For example, metal package closures in fresh-cut cabbage used for coleslaw have resulted in broken teeth. X-ray analysis of packaged goods is an effective method of detecting metallic physical hazards and should be a CCP in fresh-cut processing, with zero tolerance for the presence of such a hazard.

SIGNIFICANCE OF GAPS, GMPS, AND HACCP FOR THE AUDITOR

Before HACCP principles are applied to fresh produce production and handling operations, an auditor should determine if the facilities and practices meet the minimum criteria for GAPs and GMPs. For example, a systems approach provides a step-by-step evaluation of the production and handling system and is the proper approach for identifying potential problem areas.

Criteria addressed by GAPs or GMPs should not necessarily be included as a part of a HACCP program. For example, utilization of contaminated water for overhead irrigation or for postharvest washing is not an acceptable GAP. Therefore, it would not be necessary to consider this as a CCP for a HACCP program.

Since the quality auditor's objective is to ensure the HACCP system results in the delivery of safe food to consumers, there is little utility in attempting to distinguish between the problems associated with violation of a GAP or GMP versus a CCP in a HACCP program for fresh produce. The auditor should be acquainted, at least generally, with the recognized critical management steps in fresh produce operations.

The following are major sources of on-farm contamination:

- Soil
- Irrigation water
- Animal manure
- Inadequately composted manure
- Wild and domesticated animals
- Inadequate fieldworker hygiene
- Harvesting equipment
- Transport containers (field to packing facility)
- Wash and rinse water
- Unsanitary handling during sorting and packaging, in packing facilities, in wholesale or retail operations, and in homes
- Equipment used to soak, pack, or cut produce
- Ice
- Cooling units (hydrocoolers)
- Transport vehicles
- Improper storage conditions (temperature)
- Improper packaging
- Cross-contamination in storage, display, and preparation

The FDA publishes guidelines for the thorough treatment of potential hazards during the production and processing of fresh produce.[4] These hazards, summarized below, are what auditors should focus on.

Production Site

Auditors should note whether fields have had applications of animal manure or municipal biosolids as fertilizer; both are potential sources of microbial hazards. Biosolids also may contain heavy metals or other chemical hazards. Either type of fertilizer may be used in accordance with known GAPs. Residual pesticides in the soil also are of concern. If it is suspected that such hazards exist, the auditor may request that appropriate testing be performed.

Pesticide Use

The EPA regulates the use of pesticides in the agricultural sector. Pesticide users are obligated to read the chemical label, act appropriately, and keep records of those actions. In the production of fresh produce, this is a GAP. In the postharvest or processing sectors, this is a GMP. Failure to use agrichemicals appropriately, regardless of whether unacceptable residues exist, can result in outright rejection of the product by the FDA. Auditors who suspect that pesticide use violations exist should request appropriate residue testing.

Water

Water of inadequate quality may be a direct source of chemical or microbial hazards, or it may serve as a vehicle for spreading localized contamination at any physical location within the production and handling environments. In many fresh produce operations, water quality management may be the only legitimate CCP in a HACCP program. All other management practices are likely to be GAPs or GMPs. Discussion of water generally is divided into two categories: agricultural water and processing water.

Agricultural Water

Agricultural water includes water used for irrigation, mixing and applying pesticides and fertilizers, cooling, and frost protection. When surface groundwater (for example, that from ponds, lakes, rivers, streams, and so on) is used for agricultural operations, testing on a regular basis is appropriate to ensure that hazards are not present. In some cases, risks are easily identifiable (for example, use of pond water for overhead irrigation when livestock are feeding and drinking in proximity to the pond). In other cases, the risk is not so obvious, as in the documented example of outbreaks of Cyclospora-related illness associated with the consumption of raspberries that were sprayed with pesticide mixed from contaminated river water. Producers also should test deep wells periodically for signs of microbial or chemical contamination.

Processing Water

Postharvest uses of water include dump tanks for transferring produce from field containers or transport vehicles to the packing line, washing, rinsing, application of waxes and fungicides, cooling, and transporting product in flumes. Management of the quality of this water is essential and should be regarded as a CCP in a HACCP program for any fresh produce handling operation. Numerous sanitation treatments are approved by the EPA, but some are commodity specific. It is the responsibility of management to read product labels to determine whether a product is suitable for the intended application and to ensure the application of the sanitation treatment is effective. Examples of such treatments include, but may not be limited to, chlorine treatment (sodium hypochlorite, calcium hypochlorite, chlorine gas, and chlorine dioxide), ozone, ultraviolet (UV) light, and peroxyacetic acid. Microbiological testing of water is useful for tracking changes in water quality. Common tests for fecal indicators do not detect the presence of viruses or parasites. Managers should keep records of water quality, treatments utilized, frequency of treatment, testing method employed, and so forth, and make these records available to auditors.

Field Sanitation

GAPs specify that equipment used in the field should be periodically sanitized and frequently inspected for indications of gross contamination with soil or organic matter. Examples of such equipment include containers, such as bulk bins, baskets, buckets, or picking bags; transport vehicles, including gondolas, trucks, and trailers; harvesting tools, including hand tools as well as mechanical devices; and any other field equipment. One example of a problem area is bulk bins, which may be used during the harvest season and then stored for several months. During the storage period rodents or birds may nest in the bins. Since rodents and birds can carry disease-causing organisms, good GAPs require that bins be sanitized prior to reuse. Auditors should request records of sanitation activities for review.

Sanitary Facilities in the Field

Most state health departments have requirements for the placement of portable toilets in fields. Regulations vary, and compliance with regulations is a challenge. The more accessible the facilities, the more likely they are to be used by workers. Few fresh fruit and vegetable facilities provide hand-washing stations near portable toilets. The California strawberry industry has been very progressive in making such facilities available to workers. Disposal of waste from toilets also is of concern because the waste can splash directly onto the crop or can leach into groundwater. Auditors should become familiar with regulatory policy in their region prior to evaluating field operations.

Sanitary Facilities in Packinghouses and Processing Plants

The employer's responsibility of providing basic sanitary facilities (restrooms) in packinghouses and processing plants is governed by a variety of state health and business regulations. Auditors' primary concerns are cleanliness of facilities, frequency of cleaning, use of disinfectants during cleaning, and whether the facility is properly supplied with hot water, hand soap, a sanitary hand-drying device, and so on. One practice widely implemented by the fresh-market tomato industry is the design of restrooms that allow for privacy without the need for a door. Hand-washing stations are located just outside the restroom so managers can verify employees' hand-washing practices.

Employee Health and Hygiene

Worker hygiene is a GMP and is governed by the CFR.[5] Auditors should pay special attention to employees who have an injury or illness that poses a potential risk for contamination of product—in this case, fresh fruits or vegetables. An example of such an injury is an infected cut on the hand. Likewise, any infectious disease accompanied by diarrhea or open sores is a risk. Employers should have safety training programs that address these issues with employees. Any evidence of improper personal hygiene is a reason for an auditor to recommend that the employee be sent home to correct the situation.

Packing Facility Sanitation

Fresh fruits and vegetables typically arrive at the packing facility with at least some degree of contamination with soil and organic material. Further, fieldworkers who may have soil on their clothes or shoes often are assigned duties in the packinghouse for a portion of the day. As much of this contamination as possible should be excluded from the packing area. Facilities should be thoroughly cleaned on at least a daily basis. Pressure cleaning, preferably with hot water and an appropriate disinfectant, is recommended for packing line machinery, floors, and all other surfaces where possible. Workers must wash their hands thoroughly before beginning their work cycle. The FDA's *Guidance for Industry: Guide to Minimize Microbial Food Safety Hazards of Fresh-Cut Fruits and Vegetables*[6] provides an excellent treatment of this topic.

Fresh-Cut Processing Facilities

Management of a fresh-cut operation is much more intensive than that required for other types of fruit and vegetable packing operations. Fresh-cut products, sometimes called minimally processed products, are more perishable, more susceptible to microbial contamination, and more likely to contain physical hazards. Raw material receiving, inspection, and precleaning, and microbiological monitoring of product and environment—all of which are optional in a standard packing operation—are essential for a fresh-cut operation where HACCP is a requirement. Details of HACCP development and auditing for fresh-cut operations are beyond the scope of this section. Auditors of fresh-cut operations must acquaint themselves with the body of literature available on the subject.

Storage and Ripening Facilities

In storage and ripening facilities, the primary concerns for auditors are cleanliness of cooling coils, walls, floors, and product containers, as well as the frequent removal of decayed product. Workers or inspectors who enter these areas should meet the requirements for personal hygiene discussed previously.

Transport of Packed Product

Transport vehicles, primarily trailers, should be inspected for condition and cleanliness prior to loading. Records of materials transported in a specific trailer should be available to auditors. Some years ago, controversy erupted when it became known that garbage had been backhauled in trailers that subsequently were used for transporting food. Trailers that have physical damage, or those with ineffective cooling systems, should be removed from service.

Retailers

GMPs that apply to packing operations should be applied to retail handlers as well. Of special importance is the personal hygiene of employees who stock fresh produce.

THIRD-PARTY VERIFICATION OF GAPS, GMPS, AND HACCP IMPLEMENTATION

Some supermarkets will not accept fresh fruits and vegetables from any company that does not have third-party (independent) auditing of GAPs and GMPs. Audits may be conducted by private companies, or, in some cases, a government-sponsored program may be appropriate. Such a program gives auditors an additional opportunity to examine the records of management practices for a produce handling operation. Auditors should question management regarding the existence of third-party verification and request permission to examine such records if they exist.

CASE STUDY—THE IMPLEMENTATION OF A HACCP PROGRAM IN A FRESH-MARKET TOMATO HANDLING OPERATION

Outbreaks of *salmonellosis* were associated with consumption of fresh-market tomatoes that originated from the same packinghouse in 1990 and 1993. Analysis of environmental samples from the packinghouse did not reveal a source of salmonella contamination. However, the Centers for Disease Control and Prevention and cooperating industry associations and state agencies designed, implemented, and verified a HACCP program with special focus on GAP implementation that was proposed to serve as a model for this industry.[7]

All the potential hazard criteria mentioned in this chapter were systematically evaluated. The following three steps in the handling system were determined to be potential problem areas: sanitation of field containers, water quality management in the dump tank, and hygiene of workers that hand-sort tomatoes on the packing line. Water quality management was the sole CCP. Other factors considered included whether GAPs or GMPs were established and followed; whether the packinghouse operator gave adequate attention to all GAPs and GMPs; whether automated chlorine regulating equipment and water quality monitoring equipment were installed at the dump tank site; and whether CCPs for dump tank conditions were established and corrective actions were specified for instances of CCP deviations.

The packer also voluntarily implemented a sampling and testing program for the presence of salmonella on his tomatoes. Since the program began, all samples have tested negative for salmonella and no further illness has been associated with the packing operation. This serves as an example of the effectiveness of a proactive approach to a food safety problem.

NOTES

1. U.S. Food and Drug Administration (FDA), *Guidance for Industry: Guide to Minimize Microbial Food Safety Hazards of Fresh-Cut Fruits and Vegetables* (Washington, DC: Center for Food Safety and Applied Nutrition, 2008), http://www.fda.gov/Food/GuidanceRegulation/GuidanceDocuments RegulatoryInformation/ProducePlantProducts/ucm064458.htm.

2. *Code of Federal Regulations,* Current Good Manufacturing Practice in Manufacturing, Packing, or Holding Human Food, title 21, volume 2, part 110 (2000).
3. FDA, *Guidance.*
4. Ibid.
5. 21 CFR 110.
6. FDA, *Guidance.*
7. James W. Rushing and Frederick J. Angulo, "Implementation of a HACCP in a Commercial Tomato Packinghouse: A Model for the Industry," *Dairy, Food, and Environmental Sanitation* 16, no. 9 (1996): 549–53.

SUGGESTED READING

Anusuya Rangarajan, Elizabeth A. Bihn, Marvin P. Pritts, and Robert B. Gravani, *Food Safety Begins on the Farm: A Grower Self-Assessment of Food Safety Risks* (Ithaca, NY: Cornell University Department of Food Science, 2003), https://gaps.cornell.edu/educational-materials/food-safety-begins-farm/

Chapter 21

Retail and Food Service

INTEGRATING HACCP IN RETAIL AND FOOD SERVICE OPERATIONS

In any retail or food service establishment, the integration of HACCP principles into daily operations and management's commitment to food safety are the foundation of an effective HACCP system. Key standard operating procedures that focus on food safety should be in place for the following activities: purchasing, receiving, storage, preparation (including thawing), cooking, hot and cold holding and displaying, cooling, reheating, and serving. In retail and food service operations, the use of thermometers, prevention of contamination (for example, through water source control or cleaning and sanitizing), good employee personal hygiene, and other appropriate food handling practices are essential to the prevention of foodborne illness.

HACCP PLAN DEVELOPMENT AND IMPLEMENTATION

The first step in the development and implementation of a HACCP plan is to identify and assemble a HACCP team. At a minimum, the team should include the leader or senior manager—very often the person in charge, the chef, or the executive chef—representatives from purchasing and maintenance, the maître d' or service manager, and the kitchen steward or sanitation manager. HACCP team members should be trained in the principles of HACCP, have a thorough understanding of their establishment's operations, and have specific accountabilities within the team. A timeline should be developed for each step or principle in the development and implementation of the HACCP plan, and sufficient resources must be provided for the training of employees and ongoing maintenance of the system.

Once the HACCP team has been identified and assembled, the company's operations and its clientele should be described to assist with hazard analysis. This allows the team to define the performance capabilities of the operation and highlight any unique requirements that the plan must address. For instance, an acceptable risk level for pathogen reduction targets would differ for an immunocompromised group, such as hospital patients, compared to an immune complete group.

Hazard Analysis

Hazard analysis in the retail or food service environment encompasses the entire food system from receiving through serving. First, menus and recipes are reviewed to identify potentially hazardous foods, whether used as an ingredient or served as a distinct item. For instance, shell eggs may be served as a breakfast item or as an ingredient in a sauce.

This may seem like a daunting task when the variety of recipes that may be encountered is considered. An approach advocated by Dr. O. Peter Snyder is to group recipes into seven fundamental processes or combinations:

1. Thick, raw protein items (such as red meat, poultry, or fish), greater than 2 inches (5 cm) thick (1 inch [2.5 cm] center to surface)
2. Thin, raw protein items, less than 2 inches (5 cm) thick
3. Stocks, sauces, and stews
4. Fruits, vegetables, starches, seeds, nuts, and fungi (such as mushrooms)
5. Dough and batters
6. Hot combination dishes
7. Cold combination dishes[1]

Likewise, the National Restaurant Association Educational Foundation's ServSafe approach to HACCP provides guidelines for food recipe systems categorized as:

1. Soups, stocks, and stews
2. Meats and game meats
3. Poultry and feathered game
4. Seafood and shellfish
5. Fruits and vegetables
6. Stuffings, dressings, batters, and breadings
7. Eggs and egg dishes
8. Salads and preprepared cold foods
9. Grain dishes
10. Desserts[2]

The flow of food describes the route of food from receiving through serving and is a key element in determining where hazards may occur. This includes the major process operations of receiving, storing, preparing, cooking, serving and holding, transporting, and cooling or reheating. At each stage, the impact of employee handling, the environment surrounding the establishment, the physical facility and equipment, and process capability are described and assessed to identify any potential exposures or risks. Process capability assessment allows the team to clarify how much food can be safely prepared and served by the establishment. In California, food handler training is required by law for at least one manager of a retail operation, which includes grocery stores, retail stores, restaurants, and food service operations at sports facilities.

The FDA's *Managing Food Safety: A Manual for the Voluntary Use of HACCP Principles for Operators of Food Service and Retail Establishments* recommends assignment of menu items or groups to food process flows in one of three categories for the development of food safety management systems:

- *Process 1: Food process with no cook step.* Ready-to-eat food that is stored, prepared, and served
- *Process 2: Food preparation for same-day service.* Food that is stored, prepared, cooked, and served
- *Process 3: Complex food preparation.* Food that is stored, prepared, cooked, cooled, reheated, hot held, and served[3]

Once potentially hazardous foods and their flow have been identified, the biological, chemical, and physical hazards that can be reasonably expected to occur must be determined. Specific hazards may be intrinsic to the food itself, and data are available from multiple sources, including regulatory authorities, to identify the hazards associated with certain foods and ingredients.

Next, the facility's ability to prevent, eliminate, or otherwise reduce a hazard to an acceptable level and prevent foodborne illness is a consideration in estimating the risk associated with the hazard. As mentioned, the establishment's client profile should be considered, as certain population segments such as children and the elderly have lower resistance to foodborne illness. However, the HACCP system must be developed to protect all customers. The size and complexity of the operation, and the resulting capability to serve complex or multi-step recipes, are important considerations, in some cases leading to the decision to purchase prepared items rather than deal with a recipe that may be difficult to prepare safely. Approved suppliers with well-documented safety programs should be specified for all potentially hazardous foods. Shellfish, for instance, must only be purchased from suppliers that appear on the FDA's public health service list of certified shellfish shippers or on lists of state-approved sources.

Critical Control Points

CCPs are specified at each step in the operation where a preventive or control measure may be applied and be effective in preventing, eliminating, or reducing to an acceptable level the risk of a biological, chemical, or physical hazard. CCPs may not be appropriate at every stage of the flow of food, but they are necessary at one or more stages for potentially hazardous foods. Controls may be defined as raw food safety certification by the supplier, washed to be safe, cooked/pasteurized to be safe, or acidified to be safe.[4]

Control points and CCPs may be distinguished for different foods and preparation methods. Using fresh raw chicken as an example, the product may contain salmonella. The bacteria may be present even if the product was received at the proper temperature of less than 41°F (5°C). However, because no action can be taken at this point to prevent, eliminate, or reduce the level of the bacteria, the receiving step is a control point. Receiving procedures are followed to measure and maintain proper temperatures and to check for visual signs of time/temperature abuse or contamination. It is at the cook step that salmonella is eliminated by cooking the chicken to a minimum temperature of 165°F (74°C) for at least 15 seconds. Therefore, the cooking step is the CCP for this specific biological hazard.

CCPs for a cold potato salad in a food service establishment or supermarket deli would include formulating, mixing or preparing, cooling, and cold storage, with acidity (pH) as well as time and temperature as critical limits. In general, CCPs for biological hazards focus on time and temperature controls during cooking, hot holding, cooling, and reheating. Employee hygiene and handling practices and cross-contamination prevention are parts of the prerequisite programs (PRPs).

Critical Limits

Critical limits are defined for each CCP to specify the times, temperatures, pH, or other parameters that must be met to achieve food safety requirements. These must be measurable and may be a combination of time and temperature limits or a one-sided (minimum or maximum) limit. The data that support selection of the limit, such as the relevant Food Code[5] requirement, state or local regulation, or other scientific research data, should be documented and maintained with the HACCP plan records. In the raw chicken example provided previously, both the internal temperature of the chicken and the time maintained at this temperature are the critical limits for the cooking CCP. Critical limits should be realistic for the establishment's work environment, considering operational capabilities of the kitchen, volume of food handled, and number of employees. CCPs and critical limits should be added to each recipe and flow diagram to integrate HACCP into the operational systems of the establishment.

Monitoring Procedures

The procedure by which critical limits are monitored for compliance is defined for each CCP in the HACCP plan. Each limit that will be monitored is specified, stating who will monitor, the specific method to be utilized, and the frequency with which monitoring will occur. It is essential that the monitoring plan be realistic and achievable for normal operations and that employees are knowledgeable about the CCPs and their respective critical limits. Ideally, monitoring is performed by those people directly involved in the operation. Easy-to-use records such as time–temperature logs that clearly specify the CCP and relevant critical limits should be provided at the workstation.

Corrective Actions

If it is determined during planned monitoring, or at any other time, that the critical limit for a CCP has not been met, corrective action must be taken according to the HACCP plan. The corrective actions are described in the plan and are intended as steps for dealing with the specific area of nonconformance. Again, utilizing the chicken cooking example, if the critical limit of cooking to 165°F (74°C) or higher for at least 15 seconds is not achieved, simply continuing to cook to the specified temperature for the stated period may be the corrective action. In some cases, destruction of the product is the approved corrective action; for instance, at the hot holding CCP, cooked chicken that is held below the critical limit of 140°F (60°C) for more than four hours must be discarded.

Repeated failure to meet a critical limit may indicate a need for additional employee training or even disciplinary action or a need to review and verify the CCPs and critical limits of the current HACCP plan. Corrective actions, like critical limits, must be measurable, specific, and suitable for normal operations. It is extremely important that the appropriate corrective action be documented for each case of nonconformance with critical limits to demonstrate ongoing compliance with the HACCP plan and due diligence in the event of an investigation into a foodborne illness or complaint.

Verification Procedures

Verification procedures of the HACCP plan in a retail establishment may include daily or weekly review of monitoring logs, internal or independent third-party auditing, and, at a minimum, annual review of the plan. Calibration of critical limit monitoring instrumentation, such as thermometers or thermocouples, constitutes a verification step. Samples may be routinely tested to assure effectiveness of CCPs. Additionally, verification of the HACCP system should be conducted if there are changes in suppliers, menus, equipment, food flow, or other changes that may introduce potential hazards, or if food is linked to a foodborne illness outbreak or complaint. Employee feedback, regulatory agency alerts, supplier information, and new food safety information are also valuable tools in identifying when the HACCP system should be reviewed to ensure it is current, effective, and followed by employees.

Record-Keeping and Documentation Procedures

Record keeping and documentation should be appropriate to each element of the HACCP plan. Records that document the development of the plan and the establishment of CCPs and critical limits, along with data that validate critical limits and corrective actions, should be compiled and maintained in a single location by the HACCP team leader or designee. This also applies to verification reports, purchase specifications, approved supplier documentation, and other records that may not require daily use.

Documents that are utilized to capture monitoring data should be easy to use and maintained near the point of use, such as time–temperature logs on a clipboard kept at a salad preparation area in a supermarket deli. Corrective actions and the identification of the person(s) recording the data and/or performing the corrective action should be included. These documents should be collected daily, reviewed for completeness and any necessary follow-up, and maintained in a central location by the designated HACCP team member.

MANAGEMENT AND EMPLOYEE TRAINING

A major challenge in retail and food service establishments is employee turnover. A training program for new employees should stress food safety and the establishment's approach to HACCP. Job descriptions or task analyses should include each individual's role in implementing and monitoring food safety systems. In general, it is preferable to train individual employees on those elements of food safety and the establishment's HACCP plan that are specific to their assignment rather than requiring an overall training program on general HACCP principles. Training should be geared to the position, as well as the language and literacy level, of the employee. It should be documented, with a written program or course outline, signed attendance record, and demonstrated proficiency in the training concepts by the employee.

Training is an ongoing process, whether in response to regulatory requirements at the state or local level or simply for continuous improvement of the operation. Positive reinforcement for compliance with food safety practices, visual aids such as hand-washing signage in multiple languages, and management leadership by example are further examples of management practices that will support an effective HACCP system.

NOTES

1. O. Peter Snyder Jr., *HACCP-TQM Retail Food Operations Manual* (St. Paul, MN: Hospitality Institute of Technology and Management, 1999).
2. National Restaurant Association, *ServSafe Training's A Practical Approach to HACCP* (Chicago: The Educational Foundation of the National Restaurant Association, 2006).
3. U.S. Food and Drug Administration (FDA), *Managing Food Safety: A Manual for the Voluntary Use of HACCP Principles for Operators of Food Service and Retail Establishments* (Washington, DC: Center for Food Safety and Applied Nutrition, 2006).
4. Snyder, HACCP-TQM.
5. U.S. Food and Drug Administration (FDA), *Food Code 2013*, http://www.fda.gov/Food/GuidanceRegulation/RetailFoodProtection/FoodCode/ucm374275.htm.

Chapter 22
Animal Food

INTRODUCTION

Animal food is classified into many different categories of products, ranging from pet food, pet treats, animal feed, raw materials, ingredients, food additives, and any other articles for food for animals or articles used in food for animals. This chapter will divide the subject into two general categories: pet food and animal feed. This chapter also further describes the food type products and their historical main hazards, and provides some additional resources for auditing animal food manufacturing and distribution facilities. In the United States, the ratio between human food production and animal food production is about 50:50. The animal food industry has many similarities to the human food industry; it uses many of the same ingredients and ingredient streams. The food safety hazards from ingredients therefore have the potential to be very similar. "There is, however, a difference in the severity of the health effects of these hazards to cats, dogs, and humans. Pets tend to be very resistant to the clinical effects of infection by human food pathogens. On the other hand, they may be very sensitive to certain natural toxins or food components (for example, alkaloids, caffeine, etc.) as well as veterinary drugs and feed additives."[1] Examples of some sensitivities are thiamin deficiency in cats (through misformulated foods, especially canned food), excess vitamin D in dog food, excess copper in sheep diets, and monensin (contamination from medicated feeds) in horse diets.

One source that identifies the serious health problems in food production is the Reportable Food Registry (RFR or Registry). The RFR is an electronic reporting portal maintained by the FDA to collect instances of food that will cause serious consequences to humans or animals. The FDA started collecting data for annual reporting in 2009. Table 22.1 shows a breakdown of the animal food hazards and their current trends. The major hazards are further described in the Hazards section. As of February 2020, the most current data are shown in Table 22.1; the last annual report was issued for fiscal year 2014.

Another more up-to-date source is the list of pet food, animal feed, and animal drug recalls found on the FDA's Animal & Veterinary website (fda.gov/animal-veterinary/safety-health/recalls-withdrawals). The archive files are by year since 2007. The most recent data from 2017 through 2020 are found in Table 22.2 (pet food) and Table 22.3 (animal feed).

Table 22.1 Reportable food registry summary—animal food hazard breakdown.

Hazards	2011-12	2012-13	2013-14
Salmonella	26%	60%	33%
Nutrient imbalance	42%	20%	44%
Drug contamination	21%	13%	11%
Listeria monocytogenes	0%	0%	11%
Foreign object	5%	3%	0%

Table 22.2 Pet food recalls by hazard breakdown (86 total recalls), 2017–2020.

Biological hazards	
Salmonella	33.7%
Listeria	20.9%
E. coli	1.2%
Clostridium	1.2%
Chemical hazards	
Nutrient imbalance	19.8%
Drug contamination	10.5%
Mycotoxins	9.6%
Other chemicals	3.5%
Physical hazards	2.3%
GMP compliance	2.3%

Table 22.3 Animal feed recalls by hazard breakdown (16 total recalls), 2017-2020.

Chemical hazards	
Nutrient imbalance	50%
Mycotoxins	19%
Drug contamination	19%
Pesticide/herbicide	6%
Biological hazards	
Salmonella/listeria	6%
Physical hazards	0%
GMP compliance	0%

PET FOOD

Cat and Dog Food

Dry pet food—Includes cat and dog food typically containing 10%–12% moisture. An example of a dry pet food is a kibble. Kibbles are made by heating through an extrusion or baking process, and then drying to a water activity of less than 0.85. The extrusion and baking should render the product microbiologically safe and shelf stable. The low water activity inhibits the growth of microorganisms and mold.

Canned pet food—Includes cat and dog food typically containing 75%–78% moisture. "Thermally processed acid foods have a good record of food safety and shelf stability...Products that have the right combination of low pH, adequate levels of acetic acid, salt and sugars can be filled without a thermal process and still be safe and stable...However, if products are not properly formulated, there can be risks to food safety or shelf stability. In particular, poorly acidified foods can lead to botulism and cold filled products with inadequate acetic acid can spoil."[2]

The following regulations or codes apply to manufacturing of thermally processed low-acid canned products:

> USFDA requirements for "Thermally Processed Low-acid Foods Packaged in Hermetically Sealed Containers," as contained in 21 CFR part 113, and "Acidified Foods" in 21 CFR, part 114, "Code of Hygienic Practice for Low-Acid and "Acidified Low-Acid Canned Foods," as published by the Codex Alimentarius Commission: (CAC/RCP 23-1979), or "Code of Hygienic Practice for Aseptically Processed and Packaged Low-Acid Foods," as published by the Codex Alimentarius Commission: (CAC/RCP 40-1993).

Raw pet food—Includes fresh, frozen, and freeze-dried cat and dog food that consists primarily of meat, bones, and organs that have not been cooked (see microbial pet food hazards below). "The recent trend towards innovation in the industry for less processed and 'fresher' product concepts has led to the introduction of raw, chilled, and frozen pet foods. Given the high incidence of microbial pathogens in raw meats, it seems unlikely that products with minimal or no heat treatments can succeed without significant attention to pathogen control strategies in their manufacture. Invariably the search for shelf-stable 'fresh' product forms will lead the industry toward emerging processing technologies such as ultra-high hydrostatic pressure (UHP or HHP) pasteurization, among others."[3]

Pet treats and snacks—Includes dog biscuits, rawhide chews, jerky strips, and semi-moist treats. Jerky and treats should be heat treated and combined with pH adjustment, drying, or the addition of preservatives to make the product microbiologically safe and shelf stable. The use of salt and/or anti-microbial agents, the application of smoke, and heating the meat before drying can also contribute to inhibiting microorganisms.

Regarding jerky pet treats, from 2007 through 2015, the FDA received about 5,200 complaints of illnesses associated with jerky treats; many were imported from China. These complaints included 1,140 canine deaths. After the FDA issued a Center for Veterinary Medicine (CVM) Update about its jerky pet treats investigation, the agency received an increase in reports from the public. The most pronounced increase was in late 2013, when FDA issued its most comprehensive update, which included a "Dear Veterinarian" letter requesting specific clinical data and providing a fact sheet for pet owners. Reported cases appear to be tapering off and did not exceed 100 reports per quarter from July 2014 through December 2015.[4, 5]

Specialty pet food—"Any domesticated animal normally maintained in a cage or tank, such as, but not limited to, gerbils, hamsters, canaries, psittacine birds, mynahs, finches, tropical fish, goldfish, snakes, and turtles."[6]

PET FOOD HAZARDS

Microbiological—Salmonella and Listeria

Salmonella is the most significant microbiological hazard in pet food. For the RFR reports from 2009 to 2013, salmonella was reported as a hazard more than two times more prevalent than any other hazard reported. In fact, all the salmonella reports were for pet food; none were for animal feed (food) in this period. The primary reason for the difference is the source of protein (animal vs. plant). Secondarily, pet food has higher potential implications on human health due to the nature of human contact with different pet foods in the household (see the section on Vectors below).

Listeria, almost exclusively, is the reason for recalls in the raw pet food category. 94% of listeria pet food recalls are found in raw pet food (44%) and frozen pet food (50%). 6% of the listeria recalls—or one recall—was found in a specialty pet food (bird food). See the most recent recall items from 2017-2020 on the FDA's Animal and Veterinary website.

Nutrient Deficiency or Toxicity (Mineral and Vitamin Concentrations)

Examples of recalls include low amounts of thiamine in cat food and elevated levels of vitamin D in dog food. Cat food with low thiamine ingested over several weeks can lead to significant implications very quickly, such as curled neck, seizure, and death. Thiamine is sensitive to heat and water. Both conditions are present during the retort process of manufacturing cat food.

Excessive vitamin D in dog food is problematic because a dog's gut absorbs vitamin D in proportion with calcium, leading to kidney failure, disorders of the cardiovascular and nervous system, and death.

Chemical/Drug Contamination

Examples include the melamine pet food recall of 2007[7] and the pentobarbital dog food recall of 2017.[8]

Vectors

Examples include: (1) indirect contamination vectors via kitchen utensils, spills, bowls, sink; (2) direct and indirect contact with other pets, such as reptiles (turtles, snakes, lizards) and amphibians (frogs) in the household, and feeder rodents; and (3) direct contamination vector via contact with pet food/treats via hand contact by adults or contact via toddlers.

ANIMAL FEED

This category includes finished animal feed for the following livestock: swine, beef cattle, dairy cattle, broiler, turkey, poultry, goat, sheep, duck/geese, fish, veal, and horses (not meant for human consumption).

Types of Animal Feed

Dry animal feed includes a mixture of ingredients measured by mass.

Liquid animal feed includes products such as fat, liquid urea, molasses, and slurries that are measured by volume.

Complete feed. A nutritionally adequate feed for animals other than man; by specific formula, it is compounded to be fed as the sole ration and is capable of maintaining life and/or promoting production without any additional substance being consumed except water.[9]

Supplement. A feed used with another to improve the nutritive balance or performance of the total and intended to be:

1. Fed undiluted as a supplement to other feeds; or
2. Offered free choice with other parts of the ration separately available; or
3. Further diluted and mixed to produce a complete feed.[10]

Medicated feed. (All types of livestock whether the livestock is for human consumption or not).

1. Type A medicated articles are products with a standardized potency containing one or more new animal drugs for use in the manufacture of a medicated feed.
2. Type B medicated feed is an animal feed containing a new animal drug and intended solely for manufacture into another Type B or Type C medicated feed.
3. Type C medicated feed contains a new animal drug that may be offered as a complete feed, offered free choice, or fed as a top dressing. It is substantially a product that is a diluted Type A, Type B, or Type C with ingredients.

ANIMAL FEED INGREDIENTS AND HAZARDS

Animal products include animal proteins such as eggs and animal co-products such as lactose, animal fats and oils, dried whey, etc. The highest potential for hazards in this group is reported to be microbiological.[11]

Plant proteins includes proteins from plant products such as soybean meal, canola meal, and flaxseed meal. The highest potential for hazards in this group is reported to be microbiological[12] followed by chemical hazards. Recalls have been reported for mixing contaminated soybean oil with animal drugs.

Plant co-products include fractions of plant materials separated by food processors (for example, dried distiller's grains are a feed material produced as a by-product of the distillation process).

Minerals, vitamins, and other micro-ingredients includes a range of organic and inorganic compounds added to animal feeds to supply specific elements such as calcium, copper, iron, phosphorus, magnesium, sodium, zinc, etc.

THE ANIMAL FOOD AUDIT

Part IV, "Auditing the Food Safety System," is comprehensive and covers both human and animal food product types. It is the reference for auditing food facilities. By using risk analysis in selecting what areas to specifically target in the audit strategy, the three areas that have the highest incidences of recalls would be a good place to start: microbiological related, nutrient toxicity/deficiency, and drug contamination.

Microbial contamination has had several known sources of contamination; recontamination of product after treatment, contamination by an individual ingredient (when no kill step was involved in manufacturing), and contamination due to insufficient kill step. High-risk areas that deserve special attention include the environmental testing program (swabbing), product testing, pest control program, and the equipment/utensil cleaning programs.

Nutrient misformulation in production has two primary root causes: weighing errors and identification errors. A weighing error can occur during manual weighing if an incorrect tare is made, incorrect units are used, decimal place recorded or read, or possibly an uncalibrated measuring system is used. An incorrect identification is using the wrong material—for instance, if two or more materials are similar in description but may have largely different vitamin or mineral concentrations. The wrong material can be incorrectly used during the weighing process, or perhaps the wrong material was given an incorrect material (material number) during the incoming receiving or quality control (QC) process step. Another cause could be a misformulation by the vitamin or mineral ingredient supplier that has gone undetected; supplier qualification/auditing programs can reduce this risk. Focus audit attention on the weighing process and the documentation around that area. Performing an on-site or desk audit on the formulation and ingredients against the ingredient statement and any guarantees is an audit area that can reduce formula-related errors.

Drug contamination can occur in two primary ways: through cross-contamination of products that are run on the same equipment and from contaminated ingredients. Cross-contamination occurs when sequencing, flushing, and cleaning after feed containing a drug is followed by a feed that is not formulated with a drug. Currently, high-profile contamination cases of animal food with pentobarbital are due to use of contaminated ingredients. Pentobarbital is used to euthanize animals, especially cats, dogs, and horses, and has entered the ingredient market. Increasing focus on possible routes of pentobarbital contamination in animal food upstream through the supply chain will be needed to reduce this risk.

Prioritizing in these areas by auditing the records and practices is a strategy to conserve audit resources.

Aids in Conducting a Hazard Analysis

The previous sections on hazards should be helpful in identifying hazards. Nothing can aid conducting a hazard analysis better than your own experience and track record, but a documented resource that helps to prioritize hazards by ingredient and product categories was undertaken by the FDA CVM's *Hazard Analysis and Risk-Based Preventive Controls for Food for Animals Draft Guidance for Industry,* January 2018. In Appendix E, "Aid to Identifying Animal Food Hazards," the FDA has identified known hazards using publicly available resources. Though there is no severity or probability assigned to the identified hazards, there are categories that have no identified hazards.

The working list of hazards the FDA has identified are:

- Biological hazard: Pathogenic bacteria (for example, salmonella; *L. monocytogenes*)
- Biological hazard: Prions (for example, BSE)
- Biological hazard: Viruses or parasites
- Chemical hazard: Animal drug residue and carryover
- Chemical hazard: Economic adulterants (for example, melamine, urea)

- Chemical hazard: Environmental and industrial chemicals (for example, dioxins; PCBs)
- Chemical hazard: Heavy metals
- Chemical hazard: Natural toxins (for example mycotoxins)
- Chemical hazard: Nutrient deficiency or toxicity
- Chemical hazard: Pesticides
- Physical hazard: Physical (for example, metal, glass, plastic)

The categories in which the FDA has not found hazards "that have been reported in animal food at levels that are known to be violative, unsafe, of concern, or that have caused illness or injury" at the time the guidance was published are listed in Table 22.4.

Table 22.4 Hazards *not* found in animal food.

Category	Example products
Dried plant protein	Kelp meal, torula, dried yeast, soy protein concentrate
Color additives	Caramel, paprika, oleoresin, canthaxanthin, tagetes meal, FD&C Blue No. 1, FD&C No. 3
Enzymes	Cellulases, lipases, proteases, phytases
Natural and art. flavors	Tarragon, garlic essential oil, peppermint oil, citral, ethyl vanillin
Preservatives	Ascorbic acid, calcium sorbate, ethoxyquin, propionic acid
Stabilizers, thickeners, and emulsifiers	Apple pectin, carrageenan, gum arabic, guar gum, xanthan gum
Hydrogenated or hydrolyzed fat or oil	Hydrogenated glycerides; hydrolyzed sucrose polyesters; hydrolyzed fat, or oil, feed grade
Non-protein nitrogen	Urea, feed-grade biuret; fermented ammoniated condensed whey (note that block supplements have been recalled for excessive non-protein nitrogen)
Amino acids and related products	DL-Methionine, glycine, taurine
Vitamins	Riboflavin, vitamin D3, niacin (note that vitamin premixes have been recalled for misformulation)
Fermentation products	Presscake, direct-fed microorganisms
Molasses and molasses solubles	Molasses sourced from beet, cane, citrus, starch

NOTES

1. Pablo A. Carrion and Larry J. Thompson, *Pet Food, Food Safety Management,* Chapter 15 (Amsterdam: Elsevier, 2014).
2. Excerpts from North Dakota State University and New South Wales Food Authority.
3. Pablo A. Carrion and Larry J. Thompson, *Pet Food, Food Safety Management,* Chapter 15 (Amsterdam: Elsevier, 2014).
4. U.S. Food and Drug Administration, https://www.fda.gov/animal-veterinary/outbreaks-and-advisories/fda-investigates-animal-illnesses-linked-jerky-pet-treats
5. Ministry for Primary Industries, "Codes of practice and other documents for pet food, animal feed, supplements, and inedibles industry." https://www.mpi.govt.nz/processing/pet-food-inedibles-animal-feed-and-supplements/codes-of-practice-and-other-documents/
6. Association of American Feed Control Officials (AAFCO), http://www.aafco.org.
7. U.S. Food and Drug Administration, "Melamine Pet Food Recall of 2007," https://www.fda.gov/AnimalVeterinary/SafetyHealth/RecallsWithdrawals/ucm129575.htm
8. U.S. Food and Drug Administration, FDA/CVM Report on the Risk from Pentobarbital in Dog Food, https:www.fda.gov/AboutFDA/CentersOffices/OfficeofFoods/CVM/CVMFOIAElectronicReadingRoom/ucm129131.htm
9. Association of American Feed Control Officials (AAFCO), http://www.aafco.org.
10. Ibid.
11. M.S. Myint et al./Animal Feed Science and Technology 133 (2007) 137-148.
12. Ibid.

Part VI

Appendices

Appendix A
Hazards in Food

Table A.1 Examples of biological hazards in food.

Pathogenic gram-negative bacteria	
Salmonella spp. *Campylobacter jejuni* *Yersinia enterocolitica* *Shigella spp.* *Coxiella burnetii* *Mycobacterium bovis* *Brucella spp.* *Vibrio parahaemolyticus* *Vibrio cholerae Serogroups O1 and O139*	*Vibrio cholerae Serogroups non-O1 and non-O139* *Vibrio vulnificus* *Cronobacter (Enterobacter sakazakii) spp.* *Aeromonas hydrophila and other spp.* *Plesiomonas shigelloides* Miscellaneous bacterial enterics *Francisella tularensis*
Pathogenic *Escherichia coli* group	
Enterotoxigenic Escherichia coli (ETEC) *Enteropathogenic Escherichia coli (EPEC)*	*Enterohemorrhagic Escherichia coli (EHEC)* *Enteroinvasive Escherichia coli (EIEC)*
Gram-positive bacteria	
Clostridium perfringens *Clostridium botulinum* *Staphylococcus aureus* *Bacillus cereus* and other *Bacillus spp.*	*Listeria monocytogenes* *Streptococcus spp.* *Enterococcus*
Parasitic protozoa and worms	
Toxoplasmosis gondii *Giardia lamblia* *Entamoeba histolytica* *Cryptosporidium parvum* *Cyclospora cayetanensis* *Trichinella spp.* *Taenia spp.* *Anisakis simplex* and related worms	*Diphyllobothrium spp.* *Nanophyetus spp.* *Eustrongylides spp.* Selected amebas not linked to food or gastrointestinal illness *Ascaris lumbricoides* *Trichuris trichiura*
Viruses	
Noroviruses *Hepatitis A virus* *Hepatitis E virus*	*Rotavirus* Other viral agents
Other pathogenic agents	
Prions and transmissible spongiform encephalopathies	

(continued)

Table A.1 Examples of biological hazards in food *(continued)*.

Natural toxins	
Ciguatoxin	Gempylotoxin
Shellfish toxins (PSP, DSP, NSP, ASP, AZP)	Pyrrolizidine alkaloids
Scombrotoxin	Venomous fish
Tetrodotoxin	Grayanotoxins
Mushroom toxins	Phytohaemagglutinin
Aflatoxins	

Sources: U.S. Food and Drug Administration (FDA), *Bad Bug Book: Handbook of Foodborne Pathogenic Microorganisms and Natural Toxins,* 2nd ed., 2012, http://www.fda.gov/downloads/Food/FoodborneIllnessContaminants/UCM297627.pdf; Food and Agriculture Organization of the United Nations (FAO), *Food Quality and Standards Service Food and Nutrition Division, Food Quality and Safety Systems—A Training Manual on Food Hygiene and the Hazard Analysis and Critical Control Point (HACCP) System* (Rome: Publishing Management Group, FAO Information Division, 1998).

Table A.2 Factors affecting the growth of some foodborne pathogens.

Organism	Growth temperature (°C)	Growth pH	Growth A_w
Salmonella spp.	6.5–47	4.5–?	>0.95[1]
Clostridium botulinum A and B	10–50	4.7–9	>0.93
Clostridium botulinum nonproteolytic B	10–50	—[2]	NR[3]
Clostridium botulinum E	3.3–15–30	—[2]	>0.965
Clostridium botulinum F	4–?	—[2]	NR
Staphylococcus aureus	7–45	4.2–9.3	>0.86
Campylobacter jejuni	25–42	5.5–8	NR
Yersinia enterocolitica	1–44	4.4–9	NR
Yersinia pseudotuberculosis	5–43	—[2]	NR
Listeria monocytogenes	0–45	4.4–9.4	>0.92[4]
Vibrio cholerae O1	8–42	6–9.6	>0.95
Vibrio cholerae non-O1	—[2]	—[2]	—[2]
Vibrio parahaemolyticus	12.8–40	5–9.6	>0.94
Clostridium perfringens	10–52	5.5–8	>0.93
Bacillus cereus	10–49	4.9–9.3	>0.95
Escherichia coli	2.5–45	4.6–9.5	>0.935

(continued)

Table A.2 Factors affecting the growth of some foodborne pathogens *(continued)*.

Organism	Growth temperature (°C)	Growth pH	Growth A_w
Shigella spp.	>8–<45	?–9–11	NR
Streptococcus pyogenes	>10–<45	4.8–<9.2	NR

[1] For a genus as large as salmonella, the A_w lower limit for species growth may vary (e.g., *S. newport* = 0.941, *S. typhimurium* = 0.945).

[2] The value, though unreported, is probably close to other species of the genus.

[3] NR denotes that no reported value could be found, but for most vegetative cells, an A_w of >0.95 would be expected.

[4] Updated values from the 1996 ICMSF Microorganisms in Foods 5: Characteristics of Microbiological Pathogens.

Most values taken from E. Mitscherlich and E. H. Marth (eds.), *Microbial Survival in the Environment* (Berlin and Heidelberg: Springer-Verlag, 1984). This is a valuable, recommended reference.

Source: U.S. Food and Drug Administration (FDA), *Hazards and Controls Guide for Dairy Foods: HACCP Guidance for Processors,* Version 1.1, June 16, 2006.

Table A.3 Examples of chemical hazards in food.

Naturally occurring chemicals

- Allergens
- Mycotoxins (e.g., aflatoxin)
- Scombrotoxin (histamine)
- Ciguatoxin
- Mushroom toxins
- Shellfish toxins
  - Paralytic shellfish poisoning (PSP)
  - Diarrheic shellfish poisoning (DSP)
  - Neurotoxic shellfish poisoning (NSP)
  - Amnesic shellfish poisoning (ASP)
  - Pyrrolizidine alkaloids
- Phytohaemagglutinin (red kidney bean poisoning)
- Grayanotoxin (honey intoxication)
- Tetrodotoxin (pufferfish)

Added chemicals

- Polychlorinated biphenyls (PCBs)
- Agricultural chemicals
- Pesticides
- Fertilizers
- Antibiotics
- Growth hormones
- Prohibited substances
- Direct
- Indirect

Toxic elements and compounds

- Lead
- Zinc
- Cadmium
- Mercury
- Arsenic
- Cyanide

Food additives

- Vitamins and minerals

(continued)

Table A.3 Examples of chemical hazards in food *(continued)*.

Contaminants	
Lubricants	Paints
Cleaners	Refrigerants
Sanitizers	Water or steam treatment chemicals
Coatings	Pest control chemicals
From packaging materials	
Plasticizers	Adhesives
Vinyl chloride	Lead
Printing/coding inks	Tin

Source: Food and Agriculture Organization of the United Nations (FAO), Food Quality and Standards Service Food and Nutrition Division, *Food Quality and Safety Systems—A Training Manual on Food Hygiene and the Hazard Analysis and Critical Control Point (HACCP) System* (Rome: Publishing Management Group, FAO Information Division, 1998).

Table A.4 Examples of physical hazards in food.

Material	Injury potential	Sources
Glass	Cuts, bleeding; may require surgery to find or remove	Bottles, jars, light fixtures, utensils, gauge covers
Wood	Cuts, infection, choking; may require surgery to remove	Field sources, pallets, boxes, building materials
Stones	Choking, broken teeth	Fields, buildings
Metal	Cuts, infection; may require surgery to remove	Machinery, fields, wire, employees
Insulation	Choking	Building materials
Bone	Choking	Improper processing
Plastic	Choking, cuts, infection; may require surgery to remove	Packaging, pallets, equipment
Personal effects	Choking, cuts, broken teeth; may require surgery to remove	Employees

Source: Food and Agriculture Organization of the United Nations (FAO), Food Quality and Standards Service Food and Nutrition Division, *Food Quality and Safety Systems—A Training Manual on Food Hygiene and the Hazard Analysis and Critical Control Point (HACCP) System* (Rome: Publishing Management Group, FAO Information Division, 1998).

Appendix B
Validation

In 2008, the Codex Alimentarius Commission published its *Guidelines for the Validation of Food Safety Control Measures* (CAC/GL 69—2008). In this document, Codex formally separates the concepts of validation and verification.

Codex defines *validation* as follows: "Validation focuses on the collection and evaluation of scientific, technical and observational information to determine whether control measures are capable of achieving their specified purpose in terms of hazard control. Validation involves measuring performance against a desired food safety outcome or target, in respect of a required level of hazard control."

Verification is defined as follows: "Verification is an ongoing activity used to determine that the control measures have been implemented as intended. Verification occurs during or after operation of a control measure through a variety of activities, including observation of monitoring activities and review of records to confirm that implementation of control measures is according to design."

Monitoring is defined as follows: "Monitoring of control measures is the on-going collection of information at the step the control measure is applied. The information establishes that the measure is functioning as intended, i.e., within established limits. Monitoring activities are typically focused on 'real-time' measurements and on the performance of a specific control measure."[1]

The Codex document states that validation activities can include:

- Reference to scientific or technical literature, previous validation studies, or historical knowledge of the performance of the control measure
- Scientifically valid experimental data that demonstrate the adequacy of the control measure
- Collection of data during operating conditions in the whole food operation
- Mathematical modeling
- Surveys

NOTE

1. Codex Alimentarius Commission, *Guidelines for the Validation of Food Safety Control Measures*, CAC/GL 69—2008, 2008.

Appendix C
NACMCF Guidelines

Hazard Analysis and Critical Control Point Principles and Application Guidelines

Adopted August 14, 1997

National Advisory Committee on Microbiological Criteria for Foods

EXECUTIVE SUMMARY

The National Advisory Committee on Microbiological Criteria for Foods (Committee) reconvened a Hazard Analysis and Critical Control Point (HACCP) Working Group in 1995. The primary goal was to review the committee's November 1992 HACCP document, comparing it to current HACCP guidance prepared by the Codex Committee on Food Hygiene. Based upon its review, the committee made the HACCP principles more concise; revised and added definitions; included sections on prerequisite programs, education, and training, and implementation and maintenance of the HACCP plan; revised and provided a more detailed explanation of the application of HACCP principles; and provided an additional decision tree for identifying critical control points (CCPs).

The committee again endorses HACCP as an effective and rational means of assuring food safety from harvest to consumption. Preventing problems from occurring is the paramount goal underlying any HACCP system. Seven basic principles are employed in the development of HACCP plans that meet the stated goal. These principles include hazard analysis, CCP identification, establishing critical limits, monitoring procedures, corrective actions, verification procedures, and record keeping and documentation. Under such systems, if a deviation occurs indicating that control has been lost, the deviation is detected and appropriate steps are taken to reestablish control in a timely manner to assure that potentially hazardous products do not reach the consumer.

In the application of HACCP, the use of microbiological testing is seldom an effective means of monitoring CCPs because of the time required to obtain results. In most instances, monitoring of CCPs can best be accomplished through the use of physical and chemical tests, and through visual observations. Microbiological criteria do, however, play a role in verifying that the overall HACCP system is working.

The committee believes that the HACCP principles should be standardized to provide uniformity in training and applying the HACCP system by industry and government. In accordance with the National Academy of Sciences recommendation, the HACCP system must be developed by each food establishment and tailored to its individual product, processing, and distribution conditions.

In keeping with the committee's charge to provide recommendations to its sponsoring agencies regarding microbiological food safety issues, this document focuses on this area. The committee recognizes that in order to assure food safety, properly designed HACCP systems must also consider chemical and physical hazards in addition to other biological hazards.

For a successful HACCP program to be properly implemented, management must be committed to a HACCP approach. A commitment by management will indicate an awareness of the benefits and costs of HACCP and include education and training of employees. Benefits, in addition to enhanced assurance of food safety, are better use of resources and timely response to problems.

The committee designed this document to guide the food industry and advise its sponsoring agencies in the implementation of HACCP systems.

DEFINITIONS

CCP decision tree: A sequence of questions to assist in determining whether a control point is a CCP.

Control: (a) To manage the conditions of an operation to maintain compliance with established criteria; (b) The state where correct procedures are being followed and criteria are being met.

Control measure: Any action or activity that can be used to prevent, eliminate, or reduce a significant hazard.

Control point: Any step at which biological, chemical, or physical factors can be controlled.

Corrective action: Procedures followed when a deviation occurs.

Criterion: A requirement on which a judgement or decision can be based.

Critical control point: A step at which control can be applied and is essential to prevent or eliminate a food safety hazard or reduce it to an acceptable level.

Critical limit: A maximum and/or minimum value to which a biological, chemical, or physical parameter must be controlled at a CCP to prevent, eliminate, or reduce to an acceptable level the occurrence of a food safety hazard.

Deviation: Failure to meet a critical limit.

HACCP: A systematic approach to the identification, evaluation, and control of food safety hazards.

HACCP plan: The written document, which is based on the principles of HACCP and which delineates the procedures to be followed.

HACCP system: The result of the implementation of the HACCP plan.

HACCP team: The group of people who are responsible for developing, implementing, and maintaining the HACCP system.

Hazard: A biological, chemical, or physical agent that is reasonably likely to cause illness or injury in the absence of its control.

Hazard analysis: The process of collecting and evaluating information on hazards associated with the food under consideration to decide which are significant and must be addressed in the HACCP plan.

Monitor: To conduct a planned sequence of observations or measurements to assess whether a CCP is under control and to produce an accurate record for future use in verification.

Prerequisite programs: Procedures, including good manufacturing practices, that address operational conditions providing the foundation for the HACCP system.

Severity: The seriousness of the effect(s) of a hazard.

Step: A point, procedure, operation, or stage in the food system from primary production to final consumption.

Validation: That element of verification focused on collecting and evaluating scientific and technical information to determine if the HACCP plan, when properly implemented, will effectively control the hazards.

Verification: Those activities, other than monitoring, that determine the validity of the HACCP plan and that the system is operating according to the plan.

HACCP PRINCIPLES

HACCP is a systematic approach to the identification, evaluation, and control of food safety hazards based on the following seven principles:

Principle 1: Conduct a hazard analysis.

Principle 2: Determine the critical control points (CCPs).

Principle 3: Establish critical limits.

Principle 4: Establish monitoring procedures.

Principle 5: Establish corrective actions.

Principle 6: Establish verification procedures.

Principle 7: Establish record-keeping and documentation procedures.

GUIDELINES FOR APPLICATION OF HACCP PRINCIPLES

Introduction

HACCP is a management system in which food safety is addressed through the analysis and control of biological, chemical, and physical hazards from raw material production, procurement, and handling to manufacturing, distribution, and consumption of the finished product. For successful implementation of a HACCP plan, management must be strongly committed to the HACCP concept. A firm commitment to HACCP by top management provides company employees with a sense of the importance of producing safe food.

HACCP is designed for use in all segments of the food industry from growing, harvesting, processing, manufacturing, distributing, and merchandising to preparing food for consumption. Prerequisite programs such as current good manufacturing practices (CGMPs) are an essential foundation for the development and implementation of successful HACCP plans. Food safety systems based on the HACCP principles have been successfully applied in food processing plants, retail food stores, and food service operations. The seven principles of HACCP have been universally accepted by government agencies, trade associations, and the food industry around the world.

The following guidelines will facilitate the development and implementation of effective HACCP plans. While the specific application of HACCP to manufacturing facilities is emphasized here, these guidelines should be applied as appropriate to each segment of the food industry under consideration.

Prerequisite Programs

The production of safe food products requires that the HACCP system be built upon a solid foundation of prerequisite programs. Examples of common prerequisite programs are listed in Appendix C.A. Each segment of the food industry must provide the conditions necessary to protect food while it is under their control. This has traditionally been accomplished through the application of CGMPs. These conditions and practices are now considered to be prerequisite to the development and implementation of effective HACCP plans. Prerequisite programs (PRPs) provide the basic environmental and operating conditions that are necessary for the production of safe, wholesome food. Many of the conditions and practices are specified in federal, state, and local regulations and guidelines (for example, CGMPs and Food Code). The Codex Alimentarius General Principles of Food Hygiene describe the basic conditions and practices expected for foods intended for international trade. In addition to the requirements specified in regulations, industry often adopts policies and procedures that are specific to their operations. Many of these are proprietary. While PRPs may have an impact on the safety of a food, they also are concerned with ensuring foods are wholesome and suitable for consumption (Appendix C.A). HACCP plans are narrower in scope, being limited to ensuring food is safe to consume.

The existence and effectiveness of PRPs should be assessed during the design and implementation of each HACCP plan. All PRPs should be documented and regularly audited. PRPs are established and managed separately from the HACCP plan. Certain aspects, however, of a PRP may be incorporated into a HACCP plan. For example, many establishments have preventive maintenance procedures for processing equipment to avoid unexpected equipment failure and loss of production. During the development of a HACCP plan, the HACCP team may decide that the routine maintenance and calibration of an oven should be included in the plan as an activity of verification. This would further ensure that all the food in the oven is cooked to the minimum internal temperature that is necessary for food safety.

Education and Training

The success of a HACCP system depends on educating and training management and employees in the importance of their role in producing safe foods. This should also include information on the control of foodborne hazards related to all stages of the food chain. It is important to recognize that employees must first understand what HACCP is and then learn the skills necessary to make it function properly. Specific training activities should include working instructions and procedures that outline the tasks of employees monitoring each CCP.

Management must provide adequate time for thorough education and training. Personnel must be given the materials and equipment necessary to perform these tasks. Effective training is an important prerequisite to successful implementation of a HACCP plan.

Developing a HACCP Plan

The format of HACCP plans will vary. In many cases the plans will be product and process specific. However, some plans may use a unit operations approach. Generic HACCP plans can serve as useful guides in the development of process and product HACCP plans; however, it is essential that the unique conditions within each facility be considered during the development of all components of the HACCP plan.

In the development of a HACCP plan, five preliminary tasks need to be accomplished before the application of the HACCP principles to a specific product and process. The five preliminary tasks are given in Figure C.1.

Assemble the HACCP Team

The first task in developing a HACCP plan is to assemble a HACCP team consisting of individuals who have specific knowledge and expertise appropriate to the product and process. It is the team's responsibility to develop the HACCP plan. The team should be multidisciplinary and include individuals from areas such as engineering, production, sanitation, quality assurance, and food microbiology. The team should also include local personnel who are involved in the operation, as they are more familiar with the variability and limitations of the operation. In addition, this fosters a sense of ownership among those who must implement the plan. The HACCP team may need assistance from outside experts who are knowledgeable in the potential biological, chemical, and/or physical hazards associated with the product and the process. However, a plan that is developed totally by outside sources may be erroneous, incomplete, and lacking in support at the local level.

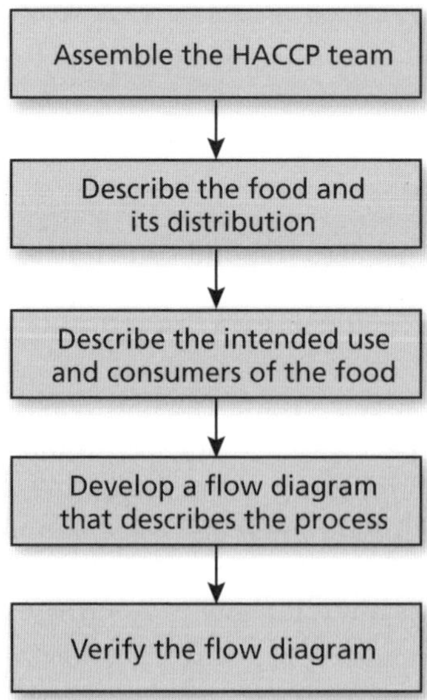

Figure C.1 Preliminary tasks in the development of the HACCP plan.

Due to the technical nature of the information required for hazard analysis, it is recommended that experts who are knowledgeable in the food process should either participate in or verify the completeness of the hazard analysis and the HACCP plan. Such individuals should have the knowledge and experience to correctly: (a) conduct a hazard analysis; (b) identify potential hazards; (c) identify hazards that must be controlled; (d) recommend controls, critical limits, and procedures for monitoring and verification; (e) recommend appropriate corrective actions when a deviation occurs; (f) recommend research related to the HACCP plan if important information is not known; and (g) validate the HACCP plan.

Describe the Food and Its Distribution

The HACCP team first describes the food. This consists of a general description of the food, ingredients, and processing methods. The method of distribution should be described along with information on whether the food is to be distributed frozen, refrigerated, or at ambient temperature.

Describe the Intended Use and Consumers of the Food

Describe the normal expected use of the food. The intended consumers may be the general public or a particular segment of the population (for example, infants, immunocompromised individuals, the elderly, etc.).

Develop a Flow Diagram That Describes the Process

The purpose of a flow diagram is to provide a clear, simple outline of the steps involved in the process. The scope of the flow diagram must cover all the steps in the process that are directly under the control of the establishment. In addition, the flow diagram can include steps in the food chain that are before and after the processing that occurs in the establishment. The flow diagram need not be as complex as engineering drawings. A block-type flow diagram is sufficiently descriptive (see Appendix C.B). Also, a simple schematic of the facility is often useful in understanding and evaluating product and process flow.

Verify the Flow Diagram

The HACCP team should perform an on-site review of the operation to verify the accuracy and completeness of the flow diagram. Modifications should be made to the flow diagram as necessary and documented.

After these five preliminary tasks have been completed, the seven principles of HACCP are applied.

Conduct a Hazard Analysis (Principle 1)

After addressing the preliminary tasks discussed previously, the HACCP team conducts a hazard analysis and identifies appropriate control measures. The purpose of the hazard analysis is to develop a list of hazards that are of such significance that they are reasonably likely to cause injury or illness if not effectively controlled. Hazards that are not reasonably likely to occur would not require further consideration within a HACCP plan. It is important to consider in the hazard analysis the ingredients and raw materials, each step in the process, product storage and distribution, and final preparation and use by the consumer.

When conducting a hazard analysis, safety concerns must be differentiated from quality concerns. A hazard is defined as a biological, chemical, or physical agent that is reasonably likely to cause illness or injury in the absence of its control. Thus, the word hazard as used in this document is limited to safety.

A thorough hazard analysis is the key to preparing an effective HACCP plan. If the hazard analysis is not done correctly and the hazards warranting control within the HACCP system are not identified, the plan will not be effective regardless of how well it is followed.

The hazard analysis and identification of associated control measures accomplish three objectives: Those hazards and associated control measures are identified. The analysis may identify needed modifications to a process or product so that product safety is further assured or improved. The analysis provides a basis for determining CCPs in Principle 2.

The process of conducting a hazard analysis involves two stages. The first, hazard identification, can be regarded as a brainstorming session. During this stage, the HACCP team reviews the ingredients used in the product, the activities conducted at each step in the process and the equipment used, the final product and its method of storage and distribution, and the intended use and consumers of the product. Based on this review, the team develops a list of potential biological, chemical, or physical hazards that may be introduced, increased, or controlled at each step in the production process. Appendix C.C lists examples of questions that may be helpful to consider when identifying potential hazards. Hazard identification focuses on developing a list of potential hazards associated with each process step under direct control of the food operation. A knowledge of any adverse health-related events historically associated with the product will be of value in this exercise.

After the list of potential hazards is assembled, stage two, the hazard evaluation, is conducted. In stage two of the hazard analysis, the HACCP team decides which potential hazards must be addressed in the HACCP plan. During this stage, each potential hazard is evaluated based on the severity of the potential hazard and its likely occurrence. Severity is the seriousness of the consequences of exposure to the hazard. Considerations of severity (for example, impact of sequelae, and magnitude and duration of illness or injury) can be helpful in understanding the public health impact of the hazard. Consideration of the likely occurrence is usually based on a combination of experience, epidemiological data, and information in the technical literature. When conducting the hazard evaluation, it is helpful to consider the likelihood of exposure and severity of the potential consequences if the hazard is not properly controlled. In addition, consideration should be given to the effects of short-term as well as long-term exposure to the potential hazard. Such considerations do not include common dietary choices that lie outside of HACCP. During the evaluation of each potential hazard, the food, its method of preparation, transportation, storage, and persons likely to consume the product should be considered to determine how each of these factors may influence the likely occurrence and severity of the hazard being controlled. The team must consider the influence of likely procedures for food preparation and storage and whether the intended consumers are susceptible to a potential hazard. However, there may be differences of opinion, even among experts, as to the likely occurrence and severity of a hazard. The HACCP team may have to rely upon the opinion of experts who assist in the development of the HACCP plan.

Hazards identified in one operation or facility may not be significant in another operation producing the same or a similar product. For example, due to differences in equipment and/or an effective maintenance program, the probability of metal contamination may be significant in one facility but not in another. A summary of the HACCP team deliberations and the rationale developed during the hazard analysis should be kept for future reference. This information will be useful during future reviews and updates of the hazard analysis and the HACCP plan.

Appendix C.D gives three examples of using a logic sequence in conducting a hazard analysis. While these examples relate to biological hazards, chemical and physical hazards are equally important to consider. Appendix C.D is for illustration purposes to further explain the stages of hazard analysis for identifying hazards. Hazard identification and evaluation as outlined in Appendix C.D may eventually be assisted by biological risk assessments as they become available. While the process and output of a risk assessment (NACMCF 1997)[1] is significantly different from a hazard analysis, the identification of hazards of concern and the hazard evaluation may be facilitated by information from risk assessments. Thus, as risk assessments addressing specific hazards or control factors become available, the HACCP team should take these into consideration.

Upon completion of the hazard analysis, the hazards associated with each step in the production of the food should be listed along with any measure(s) that are used to control the hazard(s). The term control measure is used because not all hazards can be prevented but virtually all can be controlled. More than one control measure may be required for a specific hazard. On the other hand, more than one hazard may be addressed by a specific control measure (for example, pasteurization of milk).

For example, if a HACCP team were to conduct a hazard analysis for the production of frozen cooked beef patties (Appendices C.B and C.D), enteric pathogens (for example, salmonella and verotoxin-producing E. coli) in the raw meat would be identified as hazards. Cooking is a control measure that can be used to eliminate these hazards. The following is an excerpt from a hazard analysis summary table for this product.

Step	Potential hazard(s)	Justification	Hazard to be addressed in plant? Y/N	Control measure(s)
5. Cooking	Enteric pathogens (e.g., salmonella, *verotoxigenic E. coli*)	Enteric pathogens have been associated with outbreaks of foodborne illness from undercooked ground beef	Y	Cooking

The hazard analysis summary could be presented in several different ways. One format is a table such as the one given previously. Another could be a narrative summary of the HACCP team's hazard analysis considerations and a summary table listing only the hazards and associated control measures.

Determine Critical Control Points (CCPs) (Principle 2)

A critical control point (CCP) is defined as a step at which control can be applied and is essential to prevent or eliminate a food safety hazard or reduce it to an acceptable level. The potential hazards that are reasonably likely to cause illness or injury in the absence of their control must be addressed in determining CCPs.

Complete and accurate identification of CCPs is fundamental to controlling food safety hazards. The information developed during the hazard analysis is essential for the HACCP team in identifying which steps in the process are CCPs. One strategy to facilitate the identification of each CCP is the use of a CCP decision tree (Examples of decision trees are given in Appendices C.E and C.F). Although application of the CCP decision tree can be useful in determining if a particular step is a CCP for a previously identified hazard, it is merely a tool and not a mandatory element of HACCP. A CCP decision tree is not a substitute for expert knowledge.

CCPs are located at any step where hazards can be either prevented, eliminated, or reduced to acceptable levels. Examples of CCPs may include: thermal processing, chilling, testing ingredients for chemical residues, product formulation control, and testing product for metal contaminants. CCPs must be carefully developed and documented. In addition, they must be used only for purposes of product safety. For example, a specified heat process, at a given time and temperature designed to destroy a specific microbiological pathogen, could be a CCP. Likewise, refrigeration of a precooked food to prevent hazardous microorganisms from multiplying, or the adjustment of a food to a pH necessary to prevent toxin formation could also be CCPs. Different facilities preparing similar food items can differ in the hazards identified and the steps that are CCPs. This can be due to differences in each facility's layout, equipment, selection of ingredients, processes employed, etc.

Establish Critical Limits (Principle 3)

A critical limit is a maximum and/or minimum value to which a biological, chemical, or physical parameter must be controlled at a CCP to prevent, eliminate, or reduce to an acceptable level the occurrence of a food safety hazard. A critical limit is used to distinguish between safe and unsafe operating conditions at a CCP. Critical limits should not be confused with operational limits, which are established for reasons other than food safety.

Each CCP will have one or more control measures to assure the identified hazards are prevented, eliminated, or reduced to acceptable levels. Each control measure has one or more associated critical limits. Critical limits may be based upon factors such as: temperature, time, physical dimensions, humidity, moisture level, water activity (A_w), pH, titratable acidity, salt concentration, available chlorine, viscosity, preservatives, or sensory information such as aroma and visual appearance. Critical limits must be scientifically based. For each CCP, there is at least one criterion for food safety that is to be met. An example of a criterion is a specific lethality of a cooking process such as a 5D reduction in salmonella. The critical limits and criteria for food safety may be derived from sources such as regulatory standards and guidelines, literature surveys, experimental results, and experts.

An example is the cooking of beef patties (Appendix C.B). The process should be designed to ensure the production of a safe product. The hazard analysis for cooked meat patties identified enteric pathogens (for example, *Verotoxigenic E. coli* such as *E. coli O157:H7*, and *Salmonellae*) as significant biological hazards. Furthermore, cooking is the step in the process at which control can be applied to reduce the enteric pathogens to an acceptable level. To ensure that an acceptable level is consistently achieved, accurate information is needed on the probable number of the pathogens in the raw patties, their heat resistance, the factors that influence the heating of the patties, and the area of the patty that heats the slowest. Collectively, this information forms the scientific basis for the critical limits that are established. Some of the factors that may affect the thermal destruction of enteric pathogens are listed in the following table. In this example, the HACCP team concluded that a thermal process equivalent to 155°F for 16 seconds would be necessary to assure the safety of this product. To ensure this time and temperature are attained, the HACCP team for one facility determined it would be necessary to establish critical limits for the oven temperature and humidity, belt speed (time in oven), patty thickness, and composition (for example, all beef, beef and other ingredients). Control of these factors enables the facility to produce a wide variety of cooked patties, all of which will be processed to a minimum internal temperature of 155°F for 16 seconds. In another facility, the HACCP team may conclude that the best approach is to use the internal patty temperature of 155°F and hold for 16 seconds as critical limits. In this second facility, the internal temperature and hold time of the patties are monitored at a frequency to ensure the critical limits are constantly met as they exit the oven. The example given below applies to the first facility.

Process step	CCP	Critical limits
5. Cooking	Yes	Oven temperature: ___°F Time; rate of heating and cooling (belt speed in ft./min.): ___ ft./min. Patty thickness: ___ in. Patty composition: e.g., all beef Oven humidity: ___ % RH

Establish Monitoring Procedures (Principle 4)

Monitoring is a planned sequence of observations or measurements to assess whether a CCP is under control and to produce an accurate record for future use in verification. Monitoring serves three main purposes. First, monitoring is essential to food safety management in that it facilitates tracking of the operation. If monitoring indicates there is a trend toward loss of control, then action can be taken to bring the process back into control before a deviation from a critical limit occurs. Second, monitoring is used to determine when there is loss of control and a deviation occurs at a CCP, that is, exceeding or not meeting a critical limit. When a deviation occurs, an appropriate corrective action must be taken. Third, it provides written documentation for use in verification.

An unsafe food may result if a process is not properly controlled and a deviation occurs. Because of the potentially serious consequences of a critical limit deviation, monitoring procedures must be effective. Ideally, monitoring should be continuous, which is possible with many types of physical and chemical methods. For example, the temperature and time for the scheduled thermal process of low-acid canned foods is recorded continuously on temperature recording charts. If the temperature falls below the scheduled temperature or the time is insufficient, as recorded on the chart, the product from the retort is retained and the disposition determined as in Principle 5. Likewise, pH measurement may be performed continually in fluids or by testing each batch before processing. There are many ways to monitor critical limits on a continuous or batch basis and record the data on charts. Continuous monitoring is always preferred when feasible. Monitoring equipment must be carefully calibrated for accuracy.

Assignment of the responsibility for monitoring is an important consideration for each CCP. Specific assignments will depend on the number of CCPs and control measures and the complexity of monitoring. Personnel who monitor CCPs are often associated with production (for example, line supervisors, selected line workers, and maintenance personnel) and, as required, quality control personnel. Those individuals must be trained in the monitoring technique for which they are responsible, fully understand the purpose and importance of monitoring, be unbiased in monitoring and reporting, and accurately report the results of monitoring. In addition, employees should be trained in procedures to follow when there is a trend toward loss of control so adjustments can be made in a timely manner to assure the process remains under control. The person responsible for monitoring must also immediately report a process or product that does not meet critical limits.

All records and documents associated with CCP monitoring should be dated and signed or initialed by the person doing the monitoring.

When it is not possible to monitor a CCP on a continuous basis, it is necessary to establish a monitoring frequency and procedure that will be reliable enough to indicate that the CCP is under control. Statistically designed data collection or sampling systems lend themselves to this purpose.

Most monitoring procedures need to be rapid because they relate to online, "real-time" processes, and there will not be time for lengthy analytical testing. Examples of monitoring activities include visual observations and measurement of temperature, time, pH, and moisture level.

Microbiological tests are seldom effective for monitoring due to their time-consuming nature and problems with assuring detection of contaminants. Physical and chemical measurements are often preferred because they are rapid and usually more effective for assuring control of microbiological hazards. For example, the safety of pasteurized milk is based upon measurements of time and temperature of heating rather than testing the heated milk to assure the absence of surviving pathogens.

With certain foods, processes, ingredients, or imports, there may be no alternative to microbiological testing. However, it is important to recognize that a sampling protocol that is adequate to reliably detect low levels of pathogens is seldom possible because of the large number of samples needed. This sampling limitation could result in a false sense of security by those who use an inadequate sampling protocol. In addition, there are technical limitations in many laboratory procedures for detecting and quantitating pathogens and/or their toxins.

Establish Corrective Actions (Principle 5)

The HACCP system for food safety management is designed to identify health hazards and to establish strategies to prevent, eliminate, or reduce their occurrence. However, ideal circumstances do not always prevail and deviations from established processes may occur. An important purpose of corrective actions is to prevent foods that may be hazardous from reaching consumers. Where there is a deviation from established critical limits, corrective actions are necessary. Therefore, corrective actions should include the following elements: (a) determine and correct the cause of noncompliance; (b) determine the disposition of noncompliant product; and (c) record the corrective actions that have been taken. Specific corrective actions should be developed in advance for each CCP and included in the HACCP plan. As a minimum, the HACCP plan should specify what is done when a deviation occurs, who is responsible for implementing the corrective actions, and that a record will be developed and maintained of the actions taken. Individuals who have a thorough understanding of the process, product, and HACCP plan should be assigned the responsibility for oversight of corrective actions. As appropriate, experts may be consulted to review the information available and to assist in determining disposition of noncompliant product.

Establish Verification Procedures (Principle 6)

Verification is defined as those activities, other than monitoring, that determine the validity of the HACCP plan and that the system is operating according to the plan. The NAS (1985)[2] pointed out that the major infusion of science in a HACCP system centers on proper identification of the hazards, CCPs, critical limits, and instituting proper verification procedures. These processes should take place during the development and implementation of the HACCP plans and maintenance of the HACCP system. An example of a verification schedule is given in Table C.1.

One aspect of verification is evaluating whether the facility's HACCP system is functioning according to the HACCP plan. An effective HACCP system requires little end-product testing, since sufficient validated safeguards are built in early in the process. Therefore, rather than relying on end-product testing, firms should rely on frequent reviews of their HACCP plan, verification that the HACCP plan is being correctly followed, and review of CCP monitoring and corrective action records.

Another important aspect of verification is the initial validation of the HACCP plan to determine that the plan is scientifically and technically sound, that all hazards have been identified, and that if the HACCP plan is properly implemented these hazards will be effectively controlled. Information needed to validate

Table C.1 Example of a company-established HACCP verification schedule.

Activity	Frequency	Responsibility	Reviewer
Verification activities scheduling	Yearly or upon HACCP system change	HACCP coordinator	Plant manager
Initial validation of HACCP plan	Prior to and during initial implementation of plan	Independent expert(s)[a]	HACCP team
Subsequent validation of HACCP plan	When critical limits changed, significant changes in process, equipment changed, after system failure, etc.	Independent expert(s)[a]	HACCP team
Verification of CCP monitoring as described in the plan (e.g., monitoring of patty cooking temperature)	According to HACCP plan (e.g., once per shift)	According to HACCP plan (e.g., line supervisor)	According to plan (e.g., quality control)
Review of monitoring, corrective action records to show compliance with the plan	Monthly	Quality assurance	HACCP team
Comprehensive HACCP system verification	Yearly	Independent expert(s)[a]	Plant manager

[a] Done by others than the team writing and implementing the plan. May require additional technical expertise as well as laboratory and plant test studies.

the HACCP plan often includes: (1) expert advice and scientific studies; and (2) in-plant observations, measurements, and evaluations. For example, validation of the cooking process for beef patties should include the scientific justification of the heating times and temperatures needed to obtain an appropriate destruction of pathogenic microorganisms (that is, enteric pathogens) and studies to confirm that the conditions of cooking will deliver the required time and temperature to each beef patty.

Subsequent validations are performed and documented by a HACCP team or an independent expert as needed. For example, validations are conducted when there is an unexplained system failure; a significant product, process, or packaging change occurs; or new hazards are recognized.

In addition, a periodic comprehensive verification of the HACCP system should be conducted by an unbiased, independent authority. Such authorities can be internal or external to the food operation. This should include a technical evaluation of the hazard analysis and each element of the HACCP plan as well as on-site review of all flow diagrams and appropriate records from operation of the plan. A comprehensive verification is independent of other verification procedures and must be performed to ensure that the HACCP plan is resulting in the control of the hazards. If the results of the comprehensive verification identify deficiencies, the HACCP team modifies the HACCP plan as necessary.

Verification activities are carried out by individuals within a company, third-party experts, and regulatory agencies. It is important that individuals doing verification have appropriate technical expertise to perform this function. The role of regulatory and industry in HACCP was further described by the NACMCF (1997)[3].

Examples of verification activities are included as Appendix C.G.

Establish Record-Keeping and Documentation Procedures (Principle 7)

Generally, the records maintained for the HACCP system should include the following:

1. A summary of the hazard analysis, including the rationale for determining hazards and control measures.
2. The HACCP plan
   - Listing of the HACCP team and assigned responsibilities
   - Description of the food, its distribution, intended use, and consumer
   - Verified flow diagram
   - HACCP plan summary table that includes information for:
     - Steps in the process that are CCPs
     - The hazard(s) of concern
     - Critical limits
     - Monitoring*
     - Corrective actions*
     - Verification procedures and schedule*
     - Record-keeping procedures*

   * A brief summary of positions responsible for performing the activity and the procedures and frequency should be provided.

   The following is an example of a HACCP plan summary table:

CCP	Hazards	Critical limit(s)	Monitoring	Corrective actions	Verification	Records

3. Support documentation such as validation records.
4. Records that are generated during the operation of the plan.

Examples of HACCP records are given in Appendix C.H.

IMPLEMENTATION AND MAINTENANCE OF THE HACCP PLAN

The successful implementation of a HACCP plan is facilitated by commitment from top management. The next step is to establish a plan that describes the individuals responsible for developing, implementing, and maintaining the HACCP system. Initially, the HACCP coordinator and team are selected and trained as necessary. The team is then responsible for developing the initial plan and coordinating its implementation. Product teams can be appointed to develop HACCP plans for specific products. An important aspect in developing these teams is to assure they have appropriate training. The workers who will be responsible for monitoring need to be adequately trained. Upon completion of the HACCP plan, operator procedures, forms, and procedures for monitoring and corrective action are developed. Often it is a good idea to develop a timeline for the activities involved in the initial implementation of the HACCP plan. Implementation of the HACCP system involves the continual application of the monitoring, record keeping, corrective action procedures, and other activities as described in the HACCP plan.

Maintaining an effective HACCP system depends largely on regularly scheduled verification activities. The HACCP plan should be updated and revised as needed. An important aspect of maintaining the HACCP system is to assure that all individuals involved are properly trained so they understand their role and can effectively fulfill their responsibilities.

NOTES

1. National Advisory Committee on Microbiological Criteria for Foods. *The Principles of Risk Assessment for Illness Caused by Foodborne Biological Agents.* Adopted April 4, 1997.
2. National Academy of Sciences, *An Evaluation of the Role of Microbiological Criteria for Foods and Food Ingredients* (Washington, DC: National Academy Press, 1985).
3. National Advisory Committee on Microbiological Criteria for Foods. "The Role of Regulatory Agencies and Industry in HACCP," *Int. J. Food Microbiol.* 21 (1994): 187–95.

APPENDIX C.A

Examples of Common Prerequisite Programs

The production of safe food products requires that the HACCP system be built upon a solid foundation of prerequisite programs. Each segment of the food industry must provide the conditions necessary to protect food while it is under their control. This has traditionally been accomplished through the application of current good manufacturing practices (CGMPs). These conditions and practices are now considered to be prerequisite to the development and implementation of effective HACCP plans. Prerequisite programs provide the basic environmental and operating conditions that are necessary for the production of safe, wholesome food. Common prerequisite programs may include, but are not limited to:

Facilities. The establishment should be located, constructed, and maintained according to sanitary design principles. There should be linear product flow and traffic control to minimize cross-contamination from raw to cooked materials.

Supplier control. Each facility should assure its suppliers have in place effective GMP and food safety programs. These may be the subject of continuing supplier guarantee and supplier HACCP system verification.

Specifications. There should be written specifications for all ingredients, products, and packaging materials.

Production equipment. All equipment should be constructed and installed according to sanitary design principles. Preventive maintenance and calibration schedules should be established and documented.

Cleaning and sanitation. All procedures for cleaning and sanitation of the equipment and the facility should be written and followed. A master sanitation schedule should be in place.

Personal hygiene. All employees and other persons who enter the manufacturing plant should follow the requirements for personal hygiene.

Training. All employees should receive documented training in personal hygiene, GMP, cleaning and sanitation procedures, personal safety, and their role in the HACCP program.

Chemical control. Documented procedures must be in place to assure the segregation and proper use of nonfood chemicals in the plant. These include cleaning chemicals, fumigants, and pesticides or baits used in or around the plant.

Receiving, storage, and shipping. All raw materials and products should be stored under sanitary conditions and the proper environmental conditions such as temperature and humidity to assure their safety and wholesomeness.

Traceability and recall. All raw materials and products should be lot-coded and a recall system in place so rapid and complete traces and recalls can be done when a product retrieval is necessary.

Pest control. Effective pest control programs should be in place.

Other examples of prerequisite programs might include quality assurance procedures; standard operating procedures for sanitation, processes, product formulations, and recipes; glass control; procedures for receiving, storage, and shipping; labeling; and employee food and ingredient handling practices.

APPENDIX C.B

Example of a Flow Diagram for the Production of Frozen Cooked Beef Patties

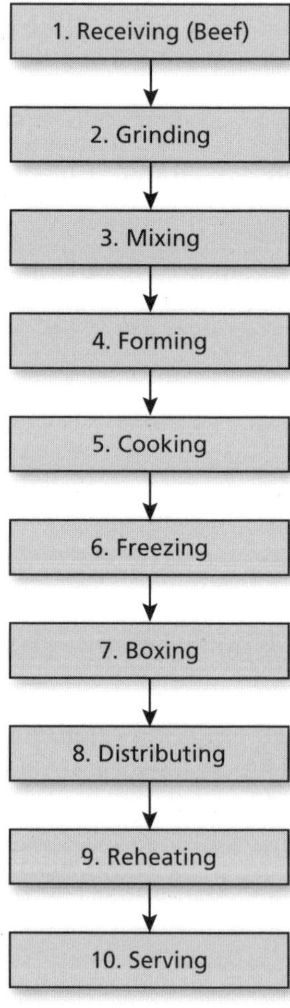

APPENDIX C.C

Examples of Questions to Be Considered When Conducting a Hazard Analysis

The hazard analysis consists of asking a series of questions that are appropriate to the process under consideration. The purpose of the questions is to assist in identifying potential hazards.

A. Ingredients

1. Does the food contain any sensitive ingredients that may present microbiological hazards (for example, salmonella, *Staphylococcus aureus*); chemical hazards (for example, aflatoxin, antibiotic or pesticide residues); or physical hazards (stones, glass, metal)?
2. Are potable water, ice, and steam used in formulating or handling the food?
3. What are the sources (for example, geographical region, specific supplier)?

B. Intrinsic factors—Physical characteristics and composition (for example, pH, type of acidulants, fermentable carbohydrate, water activity, preservatives) of the food during and after processing

1. What hazards may result if the food composition is not controlled?
2. Does the food permit survival or multiplication of pathogens and/or toxin formation in the food during processing?
3. Will the food permit survival or multiplication of pathogens and/or toxin formation during subsequent steps in the food chain?
4. Are there other similar products in the marketplace? What has been the safety record for these products? What hazards have been associated with the products?

C. Procedures used for processing

1. Does the process include a controllable processing step that destroys pathogens? If so, which pathogens? Consider both vegetative cells and spores.
2. If the product is subject to recontamination between processing (for example, cooking, pasteurizing) and packaging, which biological, chemical, or physical hazards are likely to occur?

D. Microbial content of the food

1. What is the normal microbial content of the food?
2. Does the microbial population change during the normal time the food is stored prior to consumption?
3. Does the subsequent change in microbial population alter the safety of the food?

4. Do the answers to the above questions indicate a high likelihood of certain biological hazards?

E. Facility design

1. Does the layout of the facility provide an adequate separation of raw materials from ready-to-eat (RTE) foods if this is important to food safety? If not, what hazards should be considered as possible contaminants of the RTE products?
2. Is positive air pressure maintained in product packaging areas? Is this essential for product safety?
3. Is the traffic pattern for people and moving equipment a significant source of contamination?

F. Equipment design and use

1. Will the equipment provide the time–temperature control that is necessary for safe food?
2. Is the equipment properly sized for the volume of food that will be processed?
3. Can the equipment be sufficiently controlled so the variation in performance will be within the tolerances required to produce a safe food?
4. Is the equipment reliable or is it prone to frequent breakdowns?
5. Is the equipment designed so it can be easily cleaned and sanitized?
6. Is there a chance for product contamination with hazardous substances, for example, glass?
7. What product safety devices are used to enhance consumer safety?
   - Metal detectors
   - Magnets
   - Sifters
   - Filters
   - Screens
   - Thermometers
   - Bone removal devices
   - Dud detectors
8. To what degree will normal equipment wear affect the likely occurrence of a physical hazard (for example, metal) in the product?
9. Are allergen protocols needed in using equipment for different products?

G. Packaging

1. Does the method of packaging affect the multiplication of microbial pathogens and/or the formation of toxins?
2. Is the package clearly labeled "Keep Refrigerated" if this is required for safety?
3. Does the package include instructions for the safe handling and preparation of the food by the end user?
4. Is the packaging material resistant to damage, thereby preventing the entrance of microbial contamination?
5. Are tamper-evident packaging features used?
6. Is each package and case legibly and accurately coded?
7. Does each package contain the proper label?
8. Are potential allergens in the ingredients included in the list of ingredients on the label?

H. Sanitation

1. Can sanitation have an impact on the safety of the food that is being processed?
2. Can the facility and equipment be easily cleaned and sanitized to permit the safe handling of food?
3. Is it possible to provide sanitary conditions consistently and adequately to assure safe foods?

I. Employee health, hygiene, and education

1. Can employee health or personal hygiene practices impact the safety of the food being processed?
2. Do the employees understand the process and the factors they must control to assure the preparation of safe foods?
3. Will the employees inform management of a problem that could impact the safety of food?

J. Conditions of storage between packaging and the end user

1. What is the likelihood that the food will be improperly stored at the wrong temperature?
2. Would an error in improper storage lead to a microbiologically unsafe food?

K. Intended use

1. Will the food be heated by the consumer?
2. Will there likely be leftovers?

L. Intended consumer

1. Is the food intended for the general public?
2. Is the food intended for consumption by a population with increased susceptibility to illness (for example, infants, the aged, the infirmed, immunocompromised individuals)?
3. Is the food to be used for institutional feeding or the home?

APPENDIX C.D

Examples of How the Stages of Hazard Analysis Are Used to Identify and Evaluate Hazards*

Hazard analysis stage	Frozen cooked beef patties produced in a manufacturing plant	Product containing eggs prepared for food service	Commercial frozen pre-cooked bones chicken for further processing
Stage 1: Hazard identification. Determine potential hazards associated with product	Enteric pathogens (i.e., *E. coli 0157:H7* and salmonella)	Salmonella in finished product	*Staphylococcus aureus* in finished product
Stage 2: Hazard evaluation. Assess severity of health consequences if potential hazard is not properly controlled	Epidemiological evidence indicates that these pathogens cause severe health effects including death among children and elderly. Undercooked beef patties have been linked to disease from these pathogens.	*Salmonellosis* is a foodborne infection causing a moderate to severe illness that can be caused by ingestion of only a few cells of salmonella.	Certain strains of *S. aureus* produce an enterotoxin that can cause a moderate foodborne illness.
Determine likelihood of occurrence of potential hazard if not properly controlled.	*E. coli 0157:H7* is of very low probability and moderate probability in raw meat.	Product is made with liquid eggs, which have been associated with past outbreaks of *salmonellosis*. Recent problems with *Salmonella serotype Enteritidis* in eggs cause increased concern. Probability of salmonella in raw eggs cannot be ruled out. If not effectively controlled, some consumers are likely to be exposed to salmonella from this food.	Product may be contaminated with *S. aureus* due to human handling during boning of cooked chicken. Enterotoxin capable of causing illness will only occur as *S. aureus* multiplies to about 1,000,000/g. Operating procedures during boning and subsequent freezing prevent growth of *S. aureus*; thus, the potential for enterotoxin formation is very low.

* For illustrative purposes only. The potential hazards identified may not be the only hazards associated with the products listed. The responses may be different for different establishments.

(continued)

Hazard analysis stage	Frozen cooked beef patties produced in a manufacturing plant	Product containing eggs prepared for food service	Commercial frozen pre-cooked bones chicken for further processing
Using information above, determine if this potential hazard is to be addressed in the HACCP plan.	The HACCP team decides that enteric pathogens are hazards for this product.	HACCP team determines that if the potential hazard is not properly controlled, consumption of product is likely to result in an unacceptable health risk.	The HACCP team determines that the potential for enterotoxin formation is very low. However, it is still desirable to keep the initial number of *S. aureus* organisms low. Employee practices that minimize contamination, rapid carbon dioxide freezing, and handling instructions have been adequate to control this potential hazard.
	Hazards must be addressed in the plan.	Hazards must be addressed in the plan.	Potential hazard does not need to be addressed in plan.

* For illustrative purposes only. The potential hazards identified may not be the only hazards associated with the products listed. The responses may be different for different establishments.

APPENDIX C.E

Example I of a CCP Decision Tree

Important considerations when using the decision tree:

- The decision tree is used after the hazard analysis.
- The decision tree is then used at the steps where a hazard that must be addressed in the HACCP plan has been identified.
- A subsequent step in the process may be more effective for controlling a hazard and may be the preferred CCP.
- More than one step in a process may be involved in controlling a hazard.
- More than one hazard may be controlled by a specific control measure.

Q1 Does this step involve a hazard of sufficient likelihood of occurrence and severity to warrant its control?

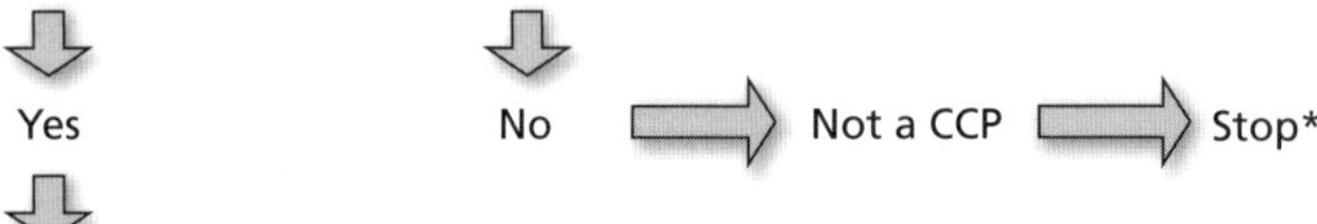

Q2 Does a control measure for the hazard exist at this step?

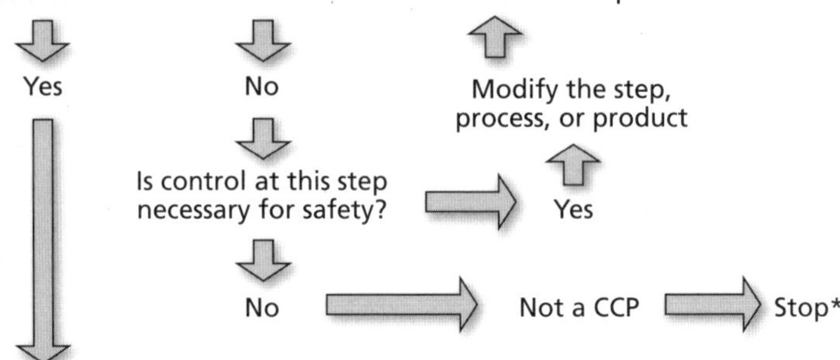

Q3 Is control at this step necessary to prevent, eliminate, or reduce the risk of the hazard to consumers?

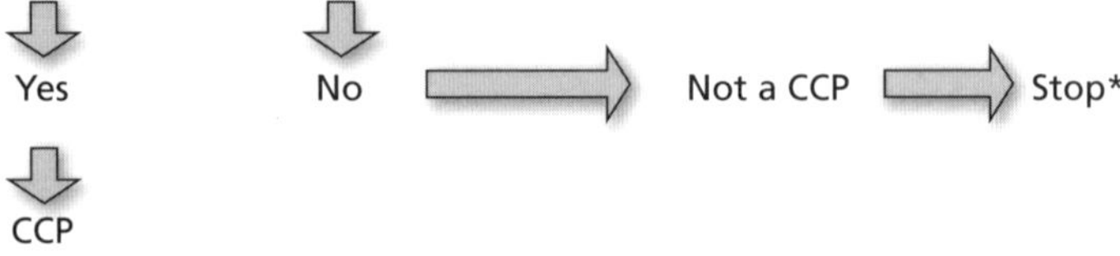

*Proceed to next step in the process.

APPENDIX C.F

Example II of a CCP Decision Tree

Q1 Do control measure(s) exist for the identified hazard?

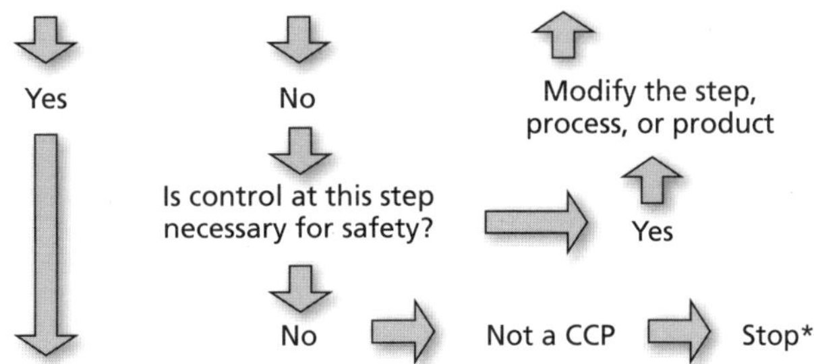

Q2 Does this step eliminate or reduce the likely occurrence of a hazard to an acceptable level?

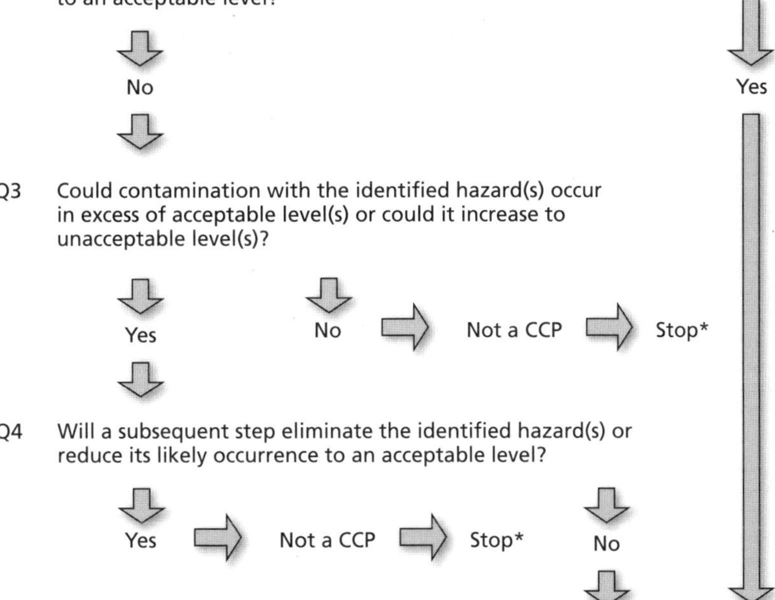

*Proceed to next step in the process.

APPENDIX C.G

Examples of Verification Activities

A. Verification procedures may include:

1. Establishment of appropriate verification schedules
2. Review of the HACCP plan for completeness
3. Confirmation of the accuracy of the flow diagram
4. Review of the HACCP system to determine if the facility is operating according to the HACCP plan
5. Review of CCP monitoring records
6. Review of records for deviations and corrective actions
7. Validation of critical limits to confirm they are adequate to control significant hazards
8. Validation of HACCP plan, including on-site review
9. Review of modifications of the HACCP plan
10. Sampling and testing to verify CCPs

B. Verification should be conducted:

1. Routinely, or on an unannounced basis, to assure CCPs are under control
2. When there are emerging concerns about the safety of the product
3. When foods have been implicated as a vehicle of foodborne disease
4. To confirm that changes have been implemented correctly after a HACCP plan has been modified
5. To assess whether a HACCP plan should be modified due to a change in the process, equipment, ingredients, etc.

C. Verification reports may include information on the presence and adequacy of:

1. The HACCP plan and the person(s) responsible for administering and updating the HACCP plan
2. The records associated with CCP monitoring
3. Direct recording of monitoring data of the CCP while in operation
4. Certification that monitoring equipment is properly calibrated and in working order
5. Corrective actions for deviations

6. Sampling and testing methods used to verify that CCPs are under control
7. Modifications to the HACCP plan
8. Training and knowledge of individuals responsible for monitoring CCPs
9. Validation activities

APPENDIX C.H

Examples of HACCP Records

A. Ingredients for which critical limits have been established
    1. Supplier certification records documenting compliance of an ingredient with a critical limit
    2. Processor audit records verifying supplier compliance
    3. Storage records (for example, time, temperature) for when ingredient storage is a CCP
B. Processing, storage, and distribution records
    1. Information that establishes the efficacy of a CCP to maintain product safety
    2. Data establishing the safe shelf life of the product, if age of product can affect safety
    3. Records indicating compliance with critical limits when packaging materials, labeling, or sealing specifications are necessary for food safety
    4. Monitoring records
    5. Verification records
C. Deviation and corrective action records
D. Employee training records that are pertinent to CCPs and the HACCP plan
E. Documentation of the adequacy of the HACCP plan from a knowledgeable HACCP expert

Appendix D
CODEX HACCP Guidelines

GUIDELINES FOR THE APPLICATION OF THE HACCP SYSTEM

Prior to application of HACCP to any sector of the food chain, that sector should be operating according to the Codex General Principles of Food Hygiene, the appropriate Codex Codes of Practice, and appropriate food safety legislation. Management commitment is necessary for implementation of an effective HACCP system. During hazard identification, evaluation, and subsequent operations in designing and applying HACCP systems, consideration must be given to the impact of raw materials, ingredients, food manufacturing practices, role of manufacturing processes to control hazards, likely end use of the product, categories of consumers of concern, and epidemiological evidence relative to food safety.

The intent of the HACCP system is to focus control at CCPs. Redesign of the operation should be considered if a hazard that must be controlled is identified, but no CCPs are found.

HACCP should be applied to each specific operation separately. CCPs identified in any given example in any Codex Code of Hygienic Practice might not be the only ones identified for a specific application or might be of a different nature.

The HACCP application should be reviewed and necessary changes made when any modification is made in the product, process, or any step.

It is important when applying HACCP to be flexible where appropriate, given the context of the application, taking into account the nature and the size of the operation.

Application

The application of HACCP principles consists of the following tasks as identified in the Logic Sequence for Application of HACCP (Figure D.1).

1. *Assemble the HACCP team.* The food operation should assure that the appropriate product-specific knowledge and expertise is available for the development of an effective HACCP plan. Optimally, this may be accomplished by assembling a multidisciplinary team. Where such expertise is not available on site, expert advice should be obtained from other sources. The scope of the HACCP plan should be identified. The scope should describe which segment of the food chain is involved and the general classes of hazards to be addressed (for example, does it cover all classes of hazards or only selected classes).

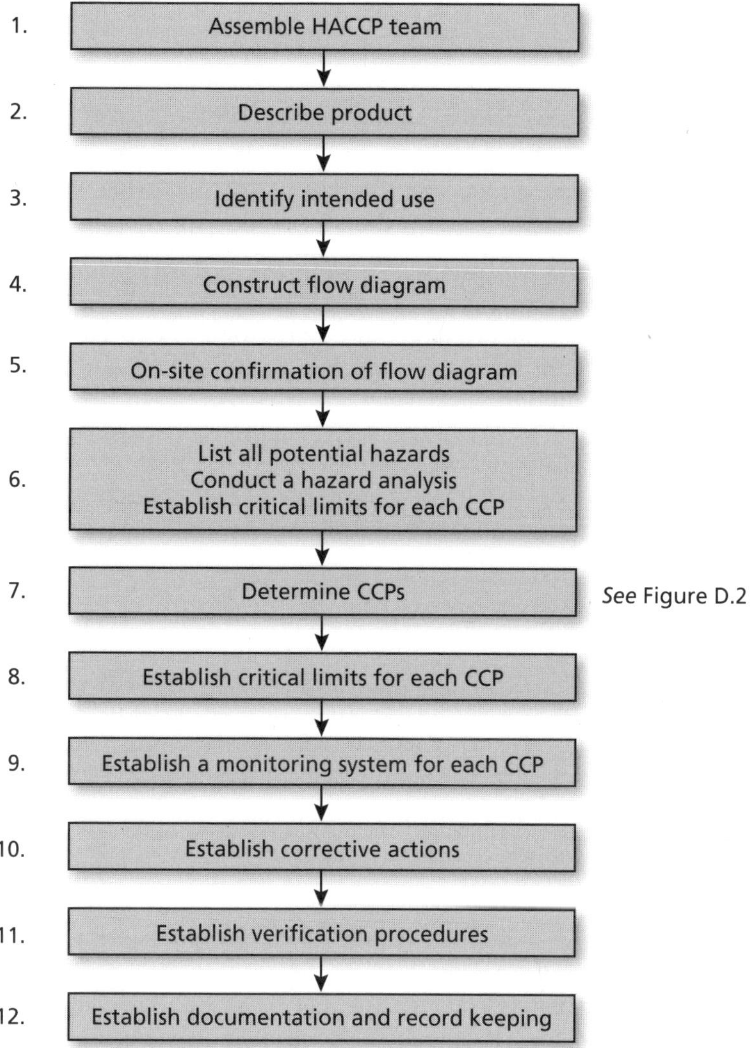

Figure D.1 Logic sequence for the application of HACCP.

2. *Describe the product.* A full description of the product should be drawn up, including relevant safety information such as: composition, physical/chemical structure (including A_w, pH, etc.), microcidal/static treatments (heat-treatment, freezing, brining, smoking, etc.), packaging, durability, and storage conditions and method of distribution.

3. *Identify intended use.* The intended use should be based on the expected uses of the product by the end user or consumer. In specific cases, vulnerable groups of the population, for example, institutional feeding, may have to be considered.

4. *Construct a flow diagram.* The flow diagram should be constructed by the HACCP team. The flow diagram should cover all steps in the operation. When applying HACCP to a given operation, consideration should be given to steps preceding and following the specified operation.

5. *On-site confirmation of flow diagram.* The HACCP team should confirm the processing operation against the flow diagram during all stages and hours of operation and amend the flow diagram where appropriate.

6. *List all potential hazards associated with each step, conduct a hazard analysis, and consider any measures to control identified hazards* (see Principle 1). The HACCP team should list all the hazards that may be reasonably expected to occur at each step, from primary production, processing, manufacture, and distribution, until the point of consumption.

The HACCP team should next conduct a hazard analysis to identify for the HACCP plan which hazards are of such a nature that their elimination or reduction to acceptable levels is essential to the production of a safe food.

In conducting the hazard analysis, wherever possible the following should be included:

- The likely occurrence of hazards and severity of their adverse health effects
- The qualitative and/or quantitative evaluation of the presence of hazards
- Survival or multiplication of microorganisms of concern
- Production or persistence in foods of toxins, chemicals, or physical agents
- Conditions leading to the above

The HACCP team must then consider what control measures, if any, exist that can be applied for each hazard.

More than one control measure may be required to control a specific hazard(s), and more than one hazard may be controlled by a specified control measure.

7. *Determine the CCPs* (see Principle 2). There may be more than one CCP at which control is applied to address the same hazard. The determination of a CCP in the HACCP system can be facilitated by the application of a decision tree (for example, Figure D.2), which indicates a logic reasoning approach. Application of a decision tree should be flexible, given whether the operation is for production, slaughter, processing, storage, distribution, or other. It should be used for guidance when determining CCPs. This example of a decision tree may not be applicable to all situations. Other approaches may be used. Training in the application of the decision tree is recommended.

If a hazard has been identified at a step where control is necessary for safety and no control measure exists at that step, or any other, then the product or process should be modified at that step, or at any earlier or later stage, to include a control measure.

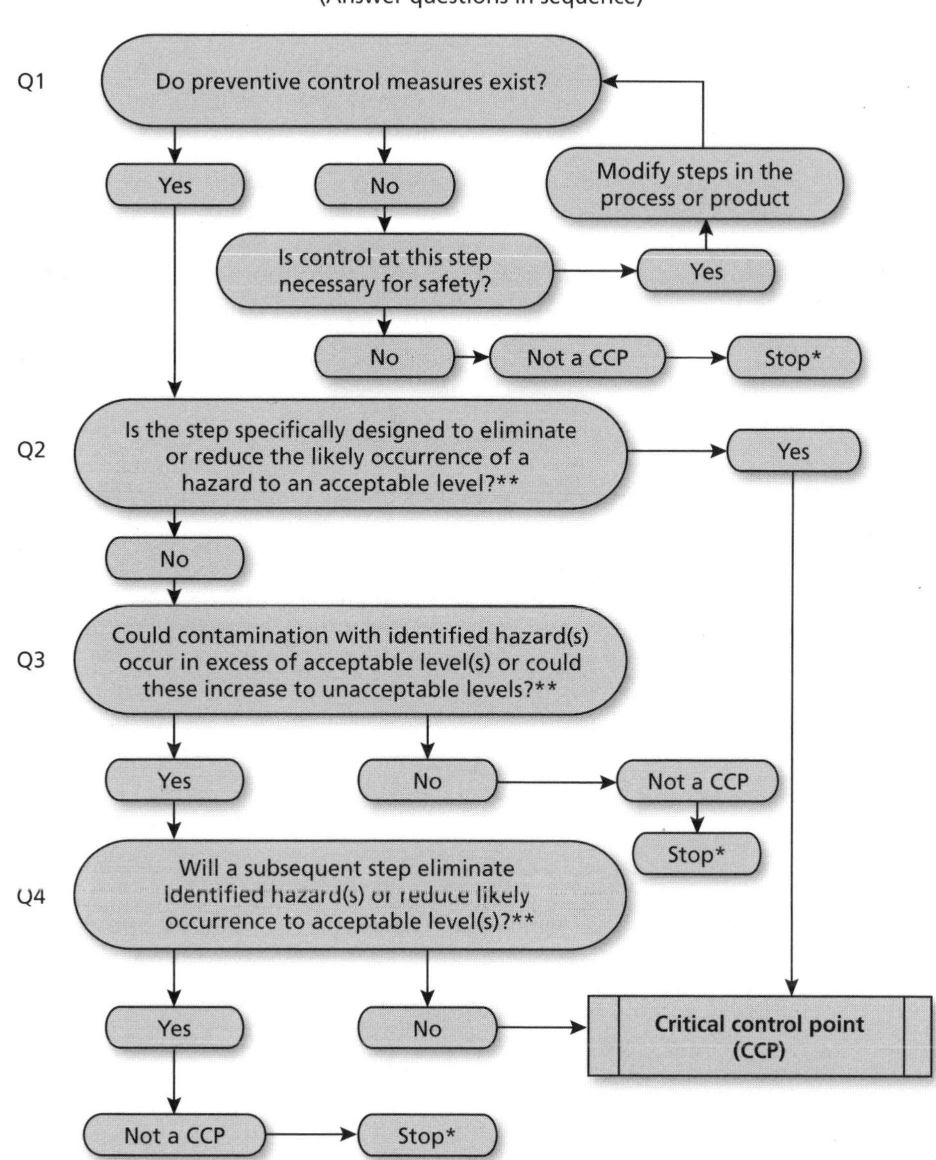

Figure D.2 Example of decision tree to identify CCPs.

8. *Establish critical limits for each CCP* (see Principle 3). Critical limits must be specified and validated if possible for each CCP. In some cases, more than one critical limit will be elaborated at a particular step. Criteria often used include measurements of temperature, time, moisture level, pH, A_w, available chlorine, and sensory parameters such as visual appearance and texture.

9. *Establish a monitoring system for each CCP* (see Principle 4). Monitoring is the scheduled measurement or observation of a CCP relative to its critical limits. The monitoring procedures must be able to detect loss of control at the CCP. Further, monitoring should ideally provide this information in time to make adjustments to ensure control of the process to prevent violating the critical limits. Where possible, process adjustments should be made when monitoring results indicate a trend toward loss of control at a CCP. The adjustments should be taken before a deviation occurs. Data derived from monitoring must be evaluated by a designated person with knowledge and authority to carry out corrective actions when indicated. If monitoring is not continuous, then the amount or frequency of monitoring must be sufficient to guarantee the CCP is in control. Most monitoring procedures for CCPs will need to be done rapidly because they relate to online processes and there will not be time for lengthy analytical testing. Physical and chemical measurements are often preferred to microbiological testing because they may be done rapidly and can often indicate the microbiological control of the product. All records and documents associated with monitoring CCPs must be signed by the person(s) doing the monitoring and by a responsible reviewing official(s) of the company.

10. *Establish corrective actions* (see Principle 5). Specific corrective actions must be developed for each CCP in the HACCP system to deal with deviations when they occur.

The actions must ensure that the CCP has been brought under control. Actions taken must also include proper disposition of the affected product. Deviation and product disposition procedures must be documented in the HACCP record keeping.

11. *Establish verification procedures* (see Principle 6). Establish procedures for verification. Verification and auditing methods, procedures, and tests, including random sampling and analysis, can be used to determine if the HACCP system is working correctly. The frequency of verification should be sufficient to confirm that the HACCP system is working effectively. Examples of verification activities include:

- Review of the HACCP system and its records
- Review of deviations and product dispositions
- Confirmation that CCPs are kept under control

Where possible, validation activities should include actions to confirm the efficacy of all elements of the HACCP plan.

12. *Establish documentation and record keeping* (see Principle 7). Efficient and accurate record keeping is essential to the application of a HACCP system. HACCP procedures should be documented. Documentation and record keeping should be appropriate to the nature and size of the operation. Documentation examples are:

- Hazard analysis
- CCP determination
- Critical limit determination

Record examples are:

- CCP monitoring activities
- Deviations and associated corrective actions
- Modifications to the HACCP system

An example of a HACCP worksheet is shown in Figure D.3.

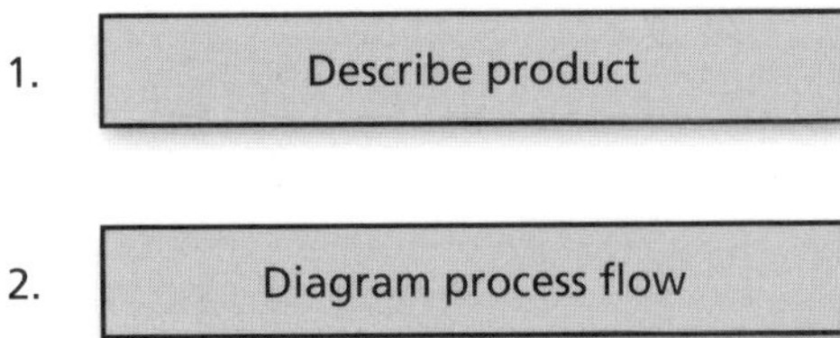

3.

List							
Step	Hazard(s)	Control measure(s)	CCPs	Critical limits	Monitoring procedures	Corrective actions	Records

4. Verification

Figure D.3 Example of a HACCP worksheet.

Training

Training of personnel in industry, government, and academia in HACCP principles and applications, and increasing awareness of consumers, are essential elements for the effective implementation of HACCP. As an aid in developing specific training to support a HACCP plan, working instructions and procedures should be developed that define the tasks of the operating personnel to be stationed at each CCP.

Cooperation between primary producer, industry, trade groups, consumer organizations, and responsible authorities is of vital importance. Opportunities should be provided for the joint training of industry and control authorities to encourage and maintain a continuous dialogue and create a climate of understanding in the practical application of HACCP.

Appendix E
ISO 22000

In September 2005, the International Organization for Standardization (ISO) published a food safety management standard titled *ISO 22000:2005, Food safety management systems—Requirements for any organization in the food chain.* This auditable standard combines all the recognized food safety elements, including HACCP principles as defined by the Codex Alimentarius Commission, prerequisite programs (PRPs), systems approach to food safety, and communications along the food chain.

ISO 22000 is written as a management system standard. As a result, the standard addresses the following elements:

- Policy
- Planning
- Implementation and operations
- Performance assessment
- Improvement
- Management review

ISO 22000 describes the requirements for a food safety management system. Therefore, it can be used to develop a food safety management system that exceeds the minimum regulatory requirements. Differences between ISO 22000 and the Codex HACCP include:

- The standard is written in auditable form. Therefore, it can be easily used to develop plans for both internal audits and third-party certification audits.
- The standard clarifies the PRPs.
- The standard links the PRPs in the HACCP plan into a single food safety management system.
- The standard requires the development of a quality policy with measurable objectives.
- The standard has requirements for management review of the food safety management system to ensure its effectiveness, suitability, and adequacy.

- There is increased responsibility for the food safety team and team leader. The food safety team is responsible for developing, implementing, and maintaining the food safety management system. The food safety team leader is responsible for reporting to top management on the effectiveness of the food safety management system.
- The organization must have effective internal and external communications with regard to food safety issues.
- The standard separates validation activities from verification activities.
- There are additional documentation requirements.

All relevant food safety hazards must be identified. Hazard analysis is then conducted on all the hazards. The organization must determine if food safety can be achieved either through operational PRPs or through the HACCP plan. The food safety control measures that are used for the operational PRPs and HACCP plan must be validated, verified, and monitored. Table E.1 provides an outline of ISO 22000.

ISO 22000 separates the PRPs into two major groups:

- *PRPs that address infrastructure and maintenance of the food safety management system.* Food-sector-specific requirements for PRPs are being defined in the ISO 22002 series of standards. As of the publication of this book, ISO has published the following standards in this series: ISO/TS 22002-1:2009, Prerequisite programs on food safety—Part 1: Food manufacturing; ISO/TS 22002-2:2013, Prerequisite programs on food safety—Part 2: Catering; and ISO/TS 22002-3:2011, Prerequisite programs on food safety—Part 3: Farming, ISO/TS2002-4: 2013. Prerequisite programs on food safety--Part 4: Food packaging manufacturing.
- *Operational prerequisite programs (OPRPs) that are used to control potential food safety hazards.* Loss of control at an OPRP could result in an economic or quality loss or a higher probability of a health risk. In many ways, OPRPs can be viewed as control points.

Thus, ISO 22000 recognizes that some food processing organizations may not be able to identify a CCP in the process.

ISO 22000 separates the concept of validation and verification. Validation is not considered to be a subset of verification. The following definitions can be used to understand the differences between validation, verification, and monitoring:

Validation—Obtaining evidence that the control measures managed by the HACCP plan and the operational PRPs are capable of being effective. This is an assessment conducted prior to starting operations.

Verification—Confirming through the provision of objective evidence that the specified requirements have been fulfilled. This is an assessment carried out during and after operations.

Table E.1 Structure of ISO 22000:2005, *Food safety management systems—Requirements for any organization in the food chain.*

Element	Description
1	Scope
2	Normative references
3	Terms and definitions
4	Food safety management system 4.1 General requirements 4.2 Documentation requirements
5	Management responsibility 5.1 Management commitment 5.2 Food safety policy 5.3 Food safety management system planning 5.4 Responsibility and authority 5.5 Food safety team leader 5.6 Communication 5.7 Emergency preparedness and response 5.8 Management review
6	Resource management 6.1 Provision of resources 6.2 Human resources 6.3 Infrastructure 6.4 Work environment
7	Planning and realization of safe products 7.1 General 7.2 Prerequisite programs (PRPs) 7.3 Preliminary steps to enable hazard analysis 7.4 Hazard analysis 7.5 Establishing the operational prerequisite programs (PRPs) 7.6 Establishing the HACCP plan 7.7 Updating of preliminary information and documents specifying the prerequisite programs and the HACCP plan 7.8 Verification planning 7.9 Traceability system 7.10 Control of nonconformity
8	Validation, verification, and improvement of the food safety management system 8.1 General 8.2 Validation of control measure combinations 8.3 Control of monitoring and measuring 8.4 Food safety management system verification 8.5 Improvement

Monitoring—Conducting a planned sequence of observations or measurements to assess whether control measures are operating as intended. This is an activity undertaken during operations.

Monitoring and verification are ongoing activities.

REFERENCE

International Organization for Standardization. *ISO 22000:2005, Food safety management systems—Requirements for any organization in the food chain.* Geneva: ISO, 2005.

Appendix F
The ASQ Code of Ethics

INTRODUCTION

The purpose of the American Society for Quality (ASQ) Code of Ethics is to establish global standards of conduct and behavior for its members, certification holders, and anyone else who may represent or be perceived to represent ASQ. In addition to the code, all applicable ASQ policies and procedures should be followed. Violations to the Code of Ethics should be reported. Differences in work style or personalities should be first addressed directly with others before escalating to an ethics issue.

The ASQ Professional Ethics and Qualifications Committee, appointed annually by the ASQ Board of Directors, is responsible for interpreting this code and applying it to specific situations, which may or may not be specifically called out in the text. Disciplinary actions will be commensurate with the seriousness of the offense and may include permanent revocation of certifications and/or expulsion from the society.

FUNDAMENTAL PRINCIPLES

ASQ requires its representatives to be honest and transparent. Avoid conflicts of interest and plagiarism. Do not harm others. Treat them with respect, dignity, and fairness. Be professional and socially responsible. Advance the role and perception of the Quality professional.

EXPECTATIONS OF A QUALITY PROFESSIONAL

1. **Act with Integrity and Honesty**
   - Strive to uphold and advance the integrity, honor, and dignity of the quality profession.
   - Be truthful and transparent in all professional interactions and activities.
   - Execute professional responsibilities and make decisions in an objective, factual, and fully informed manner.

- Accurately represent and do not mislead others regarding professional qualifications, including education, titles, affiliations, and certifications.
- Offer services, provide advice, and undertake assignments only in your areas of competence, expertise, and training.

2. **Demonstrate Responsibility, Respect, and Fairness**
   - Hold paramount the safety, health, and welfare of individuals, the public, and the environment.
   - Avoid conduct that unjustly harms or threatens the reputation of the society, its members, or the quality profession.
   - Do not intentionally cause harm to others through words or deeds. Treat others fairly, courteously, with dignity, and without prejudice or discrimination.
   - Act and conduct business in a professional and socially responsible manner.
   - Allow diversity in the opinions and personal lives of others.

3. **Safeguard Proprietary Information and Avoid Conflicts of Interest**
   - Ensure the protection and integrity of confidential information.
   - Do not use confidential information for personal gain.
   - Fully disclose and avoid any real or perceived conflicts of interest that could reasonably impair objectivity or independence in the service of clients, customers, employers, or the society.
   - Give credit where it is due.
   - Do not plagiarize. Do not use the intellectual property of others without permission. Document the permission as it is obtained.

Appendix G
ASQ Body of Knowledge

CERTIFIED FOOD SAFETY AND QUALITY AUDITOR (CFSQA) BODY OF KNOWLEDGE

The topics in this Body of Knowledge include additional detail in the form of subtext explanations and the cognitive level at which the questions will be written. This information will provide useful guidance for both the Exam Development Committee and the candidate preparing to take the exam. The subtext is not intended to limit the subject matter or be all-inclusive of what might be covered in an exam. It is meant to clarify the type of content to be included in the exam. The descriptor in parentheses at the end of each entry refers to the maximum cognitive level at which the topic will be tested. A complete description of cognitive levels is provided at the end of this document.

I. Food Safety and HACCP System (27 Questions)

A. HACCP Terminology

Define, describe, and apply basic terms and elements related to a HACCP system including: 1) deviation, 2) hazard condition, 3) validation, 4) verification, 5) National Advisory Committee on Microbiological Criteria for Foods (NACMCF), and 6) Codex Alimentarius. (Apply)

B. Food Safety Terminology

Describe and apply the connection between basic terms related to a food safety system including: 1) food safety, 2) food safety culture, 3) food quality, 4) food quality plan, and 5) animal food and animal feed. (Apply)

C. Prerequisite Programs

1. Foundations for a food safety and HACCP system

Define and describe the foundations for a food safety and HACCP system that control the operational conditions within a food establishment such as: (Analyze)

a. Good manufacturing practices (GMPs), including personal hygiene programs

b. Good agricultural practices (GAPs)
c. Good laboratory practices (GLPs), including testing continuity plan
d. Sanitation standard operating procedures (SSOPs)
e. Chemical and hazardous materials control
f. Employee training
g. Calibration of equipment
h. Integrated pest management (IPM)
i. Foreign material control (for example, wood, metal, glass, brittle plastic, and ceramic control)
j. Maintenance programs (for example, preventive, routine, emergency, and temporary)
k. Waste management
l. Supplier and material qualification (for example, raw materials, finished goods, and primary packaging)
m. Distribution and transportation

2. Product traceability and recall

   Define and distinguish between material identification and status in relation to product traceability and recall such as label control, mock recalls, and traceability exercises. (Analyze)

3. Crisis management

   Understand and apply crisis management plans including business continuity and outbreak management. (Apply)

4. Food defense and facility design

   Apply facility design, security methods, and operational conditions necessary to mitigate bioterrorism threats and intentional adulteration. (Apply)

5. Environmental control and monitoring

   Apply various programs to support proper environmental conditions such as: 1) controls for temperature, 2) humidity, 3) dust, 4) pathogens, 5) water, 6) air and ice safety, and 7) facility design elements. (Analyze)

6. Food fraud

   Understand the impact caused by the intentional or unintentional use of ingredients (for example, substitution, mislabeling, misbranding, dilution, and counterfeiting) that may compromise economic integrity, safety of the final product, or quality of the final product. (Understand)

D. Preventive Controls

1. Process controls

   Analyze appropriate procedures, practices, and processes for safe manufacturing, processing, packing, or holding of food to significantly minimize or prevent hazards, including but not limited to sanitation, process, CCP practices, and prerequisite programs. (Analyze)

2. Supply chain control

   Apply supplier preventive process control measures and methods (for example, sanitary transport, appropriate in-house storage, and appropriate labeling) used for hazard analysis and control, supplier performance and for documenting process control. (Apply)

3. Allergen control

   Analyze specifications used for control within an allergen management process (for example, storage, labeling, packaging, and shipping). Explain how specifications are used for preventing or mitigating cross-contact and cross-contamination. (Analyze)

II. Food Safety and HACCP Management (9 Questions)

A. Preliminary Tasks

Use the following preliminary tasks to develop a food safety and HACCP system. (Apply)

1. Assemble and train the food safety and HACCP team, including qualified individuals.
2. Describe the product and its distribution.
3. Describe the intended use of the product and its end user (for example, consumer, patient, vulnerable group).
4. Develop a product or process flow diagram.
5. Verify the product or process flow diagram.

B. System Scope

Define the scope of a food safety and HACCP system in terms of product-safety management. Describe how that scope affects the relationship between HACCP and other systems, such as quality management, risk management, the Global Food Safety Initiative (GFSI). Describe the impact that non-safety regulatory requirements and customer specifications can have on the scope of a food safety HACCP system. (Evaluate)

C. Management Responsibility

Understand the importance of management's commitment to food safety and HACCP prerequisite programs, preventive controls, establishment of a food safety culture, and current and emerging domestic and global standards. (Apply)

III. HACCP Principles (22 Questions)

A. Principle 1—Hazard Analysis

Conduct a hazard analysis by: 1) identifying hazards; 2) evaluating them in terms of severity and likelihood of occurrence utilizing tools such as a risk matrix; and 3) establishing control measures for any hazards that are likely to occur. (Analyze)

B. Principle 2—Critical Control Points (CCPs)

Define and distinguish between: 1) control points, and 2) critical control points (CCPs) in various operations; and then 3) develop and use CCP decision trees. (Analyze)

C. Principle 3—Critical Limits

Describe and distinguish between various types of limits, including: 1) operational and process control limits, and 2) specification limits. Identify and use appropriate scientific sources related to chemical, microbiological and physical limits, etc., as the basis for establishing critical limits. (Apply)

D. Principle 4—Monitoring

Establish monitoring procedures that include details about: 1) whether to use continuous or scheduled (intermittent) monitoring, 2) how frequently data should be gathered and by whom, and 3) what sampling and testing methods to use in support of these procedures. (Apply)

E. Principle 5—Corrective Action

Use the following steps to establish corrective action procedures. (Analyze)

1. Identify the cause of the deviation.
2. Determine disposition of affected product.
3. Identify and document corrective action.
4. Implement corrective action and determine its effectiveness.
5. Reevaluate the HACCP plan after changes have been made.

F. Principle 6—Verification

Use the following steps to establish verification procedures for ongoing assessment. (Analyze)

1. Verify prerequisites and CCPs.
2. Review documents and records.
3. Review calibration processes and system operation.
4. Test and analyze product samples.
5. Validate the HACCP system.

G. Principle 7—Record Keeping and Documentation

Establish procedures for maintaining these elements. (Apply)

1. Documents and records used to develop the initial HACCP plan
2. CCP monitoring records
3. Records of corrective actions taken in response to deviations, including root cause analysis results, verification activities, etc.
4. A formal document control system

IV. Implementation and Maintenance of Food Safety and HACCP System (21 Questions)

A. Implementation and Assessment

Use the following steps to implement the system. (Apply)

1. Conduct a pilot or initiate the system.
2. Conduct operational qualifications (CCPs, process control plans, etc.).
3. Assess training programs.
4. Evaluate the project's effectiveness in relation to its stated objectives.
5. Review the system requirements (regulatory, internal, etc.) to determine whether changes need to be made.

B. Validation and Reassessment

Use the following steps to assess an ongoing system. (Evaluate)

1. Validate the stated system objectives in relation to the results of the pilot, system initiation, or product/process change as needed.
2. Reassess the system periodically to verify that the requirements are met through reviewing data sources such as complaints, recalls, deviations, and corrective actions.

C. Verification and Maintenance

Review various food safety and HACCP system records, including: 1) monitoring, 2) corrective action, 3) calibration, 4) training and review, 5) record-keeping procedures, and

6) operational procedures when the system is active to confirm that they are being implemented properly. (Apply)

V. Auditing Fundamentals (23 Questions)

A. Basic Terms and Concepts

Define and distinguish between quality assurance and quality control. (Apply)

B. Purpose of Audits

Explain how audits can be used to assess a wide variety of activities, including: 1) organizational effectiveness, 2) system and process effectiveness, 3) performance measurement, 4) risk management, and 5) conformance to requirements. (Analyze)

C. Types of Audits

Define and distinguish between various audit types, including: 1) product, 2) process, 3) system, 4) first-, second-, and third-party, 5) compliance, etc. (Analyze)

D. Audit Criteria

Define and distinguish between various audit criteria, such as: 1) standards, 2) contracts, 3) specifications, 4) policies, and 5) regulations. (Analyze)

E. Audit Participants

Define and describe the roles and responsibilities of various audit participants, including: 1) audit team members, 2) lead auditor, 3) client, 4) auditee, and 5) technical or subject matter experts. (Apply)

F. Ethical, Legal, and Professional Issues

1. Audit credibility

   Identify and apply ethical factors that influence audit credibility such as auditor independence, objectivity, and qualifications. (Apply)

2. Liability issues

   Identify potential legal and financial ramifications of improper auditor actions (for example, carelessness and negligence) and the effects such actions can have on liability issues for all parties. (Apply)

3. Professional conduct and responsibilities

   Define and apply the concepts of due diligence and due care with respect to confidentiality, conflict of interest, the discovery of illegal activities or unsafe conditions, etc. (Apply)

VI. Auditing Process and Auditor Competencies (23 Questions)

A. Audit Preparation and Planning

1. Elements of audit planning

   Identify and implement audit planning steps, including verifying audit authority, determining the purpose, scope, type of audit, requirements to audit against, and resources necessary, such as size and number of audit teams. (Evaluate)

2. Pre-audit documents

   Identify and analyze pre-audit documents such as audit criteria or reference materials, prior audit results, etc. (Evaluate)

3. Auditing strategies

   Identify and use various tactical methods for conducting an audit, including forward- and backward-tracing, discovery, etc. (Apply)

B. Audit Performance

1. Opening meeting

   Describe the elements of an opening meeting, including explaining to the auditee the purpose, scope, and elements of the audit to be conducted. (Apply)

2. Data collection and analysis

   Select and apply various data collection methods, such as obtaining access to documents, interviewing people, observing work activities, taking physical measurements, examining paper and electronic documents, and confirming flow diagrams, and analyze the results. (Evaluate)

3. Working papers

   Identify types of working papers, such as checklists, auditor notes, attendance rosters, etc., and determine their importance in providing evidence for an audit trail. (Evaluate)

4. Objective evidence

   Identify and differentiate various characteristics of objective evidence, such as observed, measured, verified, and documented. (Analyze)

5. Observations

   Evaluate the significance of observations in terms of positive, negative, chronic, isolated, and systemic. (Evaluate)

6. Nonconformances

   Classify nonconformances in terms of significance, severity, frequency, and level of risk. (Evaluate)

7. Audit process management

   Define and apply elements of managing an audit as it is being performed, including coordinating team and team member activities, reallocating resources, adjusting audit plans when necessary, and communicating with the auditee as needed. (Analyze)

8. Exit meeting

   Describe the elements of an exit meeting, including presenting audit observations and findings to the auditee and discussing post-audit activities, who will be responsible for performing them, and their deadlines. (Apply)

C. Audit Reporting

1. Basic steps

   Implement the common steps in generating an audit report, including reviewing and finalizing results, organizing and summarizing details, obtaining necessary approvals for report distribution, etc. (Evaluate)

2. Effective audit reports

   Evaluate various components that make audit reports effective: for example, executive summary, prioritized data, graphical data presentation, and the impact of conclusions. (Evaluate)

D. Audit Follow-up and Closure

1. Corrective and preventive action (CAPA)

   Identify and apply CAPA elements, including problem identification, assigning responsibility, root cause analysis, recurrence prevention, etc. (Apply)

2. Review and verification of corrective action plans

   Use various methods to verify and evaluate corrective actions plans, including examining revised procedures and processes or re-auditing to confirm the adequacy of corrective actions taken. (Apply)

3. Follow-up on ineffective corrective actions

   Identify and develop strategies to use when corrective actions are not implemented or are not effective, including communicating to the next level of management, re-issuing the corrective action, re-auditing, etc. (Evaluate)

4. Audit closure

   Identify various elements of audit closure and any criteria that have not been met and would prevent an audit from being closed. (Evaluate)

5. Records retention

   Identify and apply record retention requirements, such as type of documents to be retained, length of time to keep them, and storage considerations. (Apply)

E. Auditor Competencies

1. Characteristics

   Identify characteristics that make auditors effective, such as interpersonal skills, problem-solving skills, close attention to detail, the ability to work independently and in a group or on a team. (Apply)

2. Conflict resolution

   Identify typical conflict situations (disagreements, auditee delaying tactics, interruptions, etc.) and determine appropriate techniques (negotiation, cool-down periods, etc.) for resolving them. (Apply)

3. Written communication techniques

   Develop and review technical reports for critical factors, including whether the document meets the needs of the intended audience, how the report will be used, what type of photographs, illustrations, or graphics will be effective, etc. (Apply)

4. Interviewing techniques

   Define and use appropriate interviewing techniques, including active listening, open-ended or closed question types, determining the significance of pauses and their length, prompting a response, clarifying by paraphrasing, etc., in various situations, such as when supervisors are present, during group interviews, a group of workers, when using a translator, etc. (Apply)

5. Team dynamics and facilitation skills

   Define and use various techniques to support team-building efforts and to help maintain group focus, both as a participant and as a team leader. Describe the classic stages of team development (forming, storming, norming, performing and adjourning, and use coaching, guidance, and other facilitation techniques to support effective teams. (Apply)

F. International Regulations and Inspections

Identify regulatory and international food sector requirements such as Food Safety Modernization Act (FSMA), FDA 21 CFR 117 and FDA 21 CFR 507, Foreign Supplier Verification Program (FSVP), FDA 9 CFR 416 and FDA 9 CFR 417, and Dietary Supplement CGMP Requirements. (Remember)

G. Auditing Schemes

Distinguish between various auditing schemes and auditee requirements including SQF Food Safety Code, FSSC 22000 Standard, BRC Global Standards, Primus, Global G.A.P., GRMS Global Red Meat Standard, IFS International Food Standard, Canada G.A.P., and Global Aquaculture Alliance. (Remember)

VII. Quality Tools and Techniques (10 Questions)

A. Basic Quality Tools

Identify, interpret, and apply the seven basic quality tools: 1) Pareto charts, 2) cause and effect diagrams, 3) flowcharts, 4) control charts, 5) check sheets, 6) scatter diagrams, and 7) histograms. (Apply)

B. Descriptive Statistics

Identify, interpret, and use: 1) measures of central tendency (mean, median, mode), and 2) dispersion (standard deviation, variance, and frequency distribution). (Apply)

C. Sampling Methods

Identify, interpret, and use sampling methods such as: 1) acceptance, 2) random, 3) stratified, and 4) define terms such as consumer and producer risk, confidence level, etc. (Analyze)

D. Statistical Process Control

Interpret the data presented in statistical process control results. (Understand)

[NOTE: this topic will be tested at the understand level; no calculations will be required.]

E. Process Capability

Identify and distinguish the basic elements of C_p and C_{pk}. (Remember)

[NOTE: this topic will be tested at the definition level; no calculations will be required.]

F. Qualitative/Quantitative Analysis and Attributes/Variables Data

Describe and distinguish between: 1) qualitative and quantitative analyses, and 2) attributes and variables data. (Apply)

LEVELS OF COGNITION BASED ON *BLOOM'S TAXONOMY*—REVISED (2001)

In addition to content specifics, the subtext for each topic in this body of knowledge (BOK) also indicates the intended complexity level of the test questions for that topic. These levels are based on "Levels of Cognition" (from *Bloom's Taxonomy*—Revised, 2001) and are presented below in rank order, from least complex to most complex.

Remember

Recall or recognize terms, definitions, facts, ideas, materials, patterns, sequences, methods, principles, etc.

Understand

Read and understand descriptions, communications, reports, tables, diagrams, directions, regulations, etc.

Apply

Know when and how to use ideas, procedures, methods, formulas, principles, theories, etc.

Analyze

Break down information into its constituent parts and recognize their relationship to one another and how they are organized; identify sublevel factors or salient data from a complex scenario.

Evaluate

Make judgments about the value of proposed ideas, solutions, etc., by comparing the proposal to specific criteria or standards.

Create

Put parts or elements together in such a way as to reveal a pattern or structure not clearly there before; identify which data or information from a complex set is appropriate to examine further or from which supported conclusions can be drawn.

Glossary

AFDO—Association of Food and Drug Officials

audit—A planned, systematic, independent, and documented examination and review to determine whether activities and related results comply with planned arrangements and whether these arrangements are implemented effectively and suitable to achieve the objectives[1*]

calibration—The comparison of a measurement instrument or system of unverified accuracy to a measurement instrument of known accuracy to detect any variation from the true value[1]

CAPA—corrective action/preventive action

CAR—corrective action request

CCP—*See* critical control point

CFR—Code of Federal Regulations

CIP—clean in place

cleaning—The removal of soil, food residue, dirt, grease, and other objectionable material[2]

COA—certificate of analysis

Codex Alimentarius Commission—An agency of FAO and WHO

contamination—The introduction or occurrence of a contaminant in food or a food environment[2]

control—(a) To manage the conditions of an operation to maintain compliance with established criteria; (b) The state where correct procedures are being followed and the criteria are being met[3]

control measure—Any action or activity that can be used to prevent, eliminate, or reduce a significant hazard[3]

control point—Any step at which biological, chemical, or physical factors can be controlled[3]; medical device definition—Any step at which a specification or quality factor(s) can be controlled[4*]

COP—clean-out-of-place

critical control point (CCP)—Any step at which control can be applied to prevent, eliminate the occurrence of, or reduce to an acceptable level a product safety hazard[3*]

critical limit (food safety)—A maximum and/or minimum value to which a biological, chemical, or physical parameter must be controlled at a CCP to prevent, eliminate, or reduce to an acceptable level the occurrence of a food safety hazard[3]

critical limit (medical device)—A maximum and/or minimum value to which a product, process, or quality parameter must be controlled at an ECP to prevent, eliminate, or reduce to an acceptable level the occurrence of a device hazard[4]

cross-functional team—A group consisting of members from more than one department that is organized to accomplish a project[1]

customer—Recipient of a product or service provided by a supplier[1]

deviation—Failure to meet a critical limit[3]

dry labbing—Recording a value in the absence of an observation

essential control point (ECP)—A term used for medical device HACCP that is similar to a CCP—a point, step, or procedure at which control can be applied and is essential to prevent, eliminate, or reduce a hazard to an acceptable level[4]

FAO—Food and Agriculture Organization, an agency of the United Nations

FDA—U.S. Food and Drug Administration, an agency of the Department of Health and Human Services

FSIS—Food Safety and Inspection Service, an agency of USDA

FSMA—Food Safety Modernization Act

GAP—good agricultural practice

GFSI—Global Food Safety Initiative

GMP—good manufacturing practice

HACCP—Hazard analysis critical control point; a systematic approach to the identification, evaluation, and control of product safety hazards[3]

hazard—A biological, chemical, or physical agent that is reasonably likely to cause illness or injury in the absence of its control[3]

hazard analysis—The process of collecting and evaluating information on hazards associated with food under consideration to decide which are significant and must be addressed in the HACCP plan[3]

HVAC—heating, ventilation, and air conditioning

ISO—International Organization for Standardization

MPI—manufacturing process instruction

MSDS—material safety data sheet

NACMCF—National Advisory Committee on Microbiological Criteria for Foods

OPRP—operational prerequisite program

pathological—Being able to develop a disease state

PCO—pest control operator

prerequisite program (PRP)—Procedures including GMPs that address operational conditions providing the foundation for the HACCP system[3]

principles of HACCP[3]

Principle 1—Conduct a hazard analysis

Principle 2—Determine the CCPs

Principle 3—Establish critical limits

Principle 4—Establish monitoring procedures

Principle 5—Establish corrective actions

Principle 6—Establish verification procedures

Principle 7—Establish record-keeping and documentation procedures

process—A set of interrelated resources and activities that transform inputs into outputs[5]

product safety system—The organization structure, procedures, processes, and resources needed to implement product safety management

PRP—prerequisite program

QSIT—quality system inspection technique

record—Document or electronic medium that furnishes objective evidence of activities performed or results achieved[1]

sanitation—The reduction by means of chemical agents or physical methods of the number of microorganisms in the environment to a level that does not compromise food safety or suitability[2*]

severity—The seriousness of the effect(s) of a hazard[3]

SSOP—sanitation standard operating procedure

supplier—Any provider whose goods and services may be used at any stage in the production, design, delivery, and use of another company's products and services[1]

TQM—total quality management

training—Refers to the skills that employees need to learn to perform or improve the performance of their current job or tasks, or the process of providing those skills[1]

USDA—U.S. Department of Agriculture

validation—That element of verification focused on collecting and evaluating scientific and technical information to determine if the HACCP plan, when properly implemented, will effectively control the hazards[3]

verification—Those activities, other than monitoring, that determine the validity of the HACCP plan and whether the system is operating according to the plan[3]

WHO—World Health Organization, an agency of the United Nations

NOTES

1. Duke Okes and Russell T. Westcott (eds.), *The Certified Quality Manager Handbook*, 2nd ed. (Milwaukee, WI: ASQ Quality Press, 2001).
2. Codex Alimentarius Commission. *Hazard Analysis and Critical Control Point (HACCP) System and Guidelines for Its Application.* Alinorm 97/13A. Rome: Codex Alimentarius Committee on Food Hygiene, 1997.
3. National Advisory Committee on Microbiological Criteria for Foods (NACMCF). *Hazard Analysis and Critical Control Point Principles and Application Guidelines.* Washington, DC: US Food and Drug Administration. August 14, 1997.
4. Association of Food and Drug Officials (AFDO). *Medical Device HACCP Training Curriculum.* Draft ed. York, PA: AFDO, 1999.
5. ANSI/ISO/ASQ Q9000:2005, *Quality management systems—Fundamentals and vocabulary* (Milwaukee, WI: American Society for Quality, 2005).

---

* Indicates that the definition has been adapted from the definition as cited in the original source.

Bibliography

AIB International. *AIB Consolidated Standards for Food Safety*. Manhattan, KS: AIB International, 1995.

———. "HACCP Overview." Presented in *HACCP Workshop*. Manhattan, KS: AIB International, 2000.

American National Standard. *ANSI/ASQC Q19011s Guidelines for Quality and/or Environmental Management Systems Auditing*. Milwaukee, WI: ASQC Quality Press, 1994.

Association of Food and Drug Officials (AFDO). *Medical Device HACCP Training Curriculum*. Draft ed. York, PA: AFDO, 1999.

———. *Seafood HACCP Training Curriculum*. 3rd ed. York, PA: AFDO, 1999.

Baking Industry Sanitation Standards Committee (BISSC). *1998 Sanitation Standards for the Design and Construction of Bakery Equipment*. Chicago: BISSC, 1998.

Bennet, William L., and Leonard L. Steed. "An Integrated Approach to Food Safety." *Quality Progress* (February 1999): 37–42.

Boyle, Elizabeth A. E., and J. K. Kelly. Getty. *Developing and Implementing HACCP in Meat Plants*. Workshop notebook compiled at Kansas State University, Manhattan, KS, 1998.

Burson, Dennis, and Mindy Brashears. *Implementing Your Company's HACCP Plan: Meat Slaughter and Processing*. Workshop notebook compiled at the University of Nebraska, Lincoln, NE, 1998.

Campden & Chorleywood Food Research Association (CCFRA). *HACCP: A Practical Guide*. 3rd ed. Technical Manual No. 42. Gloucestershire, UK, 2003.

Canadian Food Inspection Agency. *Food Safety Enhancement Program Implementation Manual*. Vol. II. Ottawa: Government of Canada, 1995.

Codex Alimentarius Commission. *Guidelines for the Application of the Hazard Analysis and Critical Control Point (HACCP) System*. Adopted by the Joint FAO/WHO Codex Commission, 20th Session, 1993.

———. *Hazard Analysis and Critical Control Point (HACCP) System and Guidelines for Its Application*. Alinorm 97/13A. Rome: Codex Alimentarius Committee on Food Hygiene, 1997.

Corlett, Donald A., Jr. *HACCP User's Manual*. Gaithersburg, MD: Aspen, 1998.

Dillon, Mike, and Chris Griffith. *How to HACCP*. N.E. Lincolnshire, UK: MD Associates, 1996.

Doyle, Michael P., Larry R. Beuchat, and Thomas J. Montville. *Food Microbiology: Fundamentals and Frontiers.* Washington, DC: ASM Press, 1997.

European Committee for Standardization. *Food Processing Machinery—Basic Concepts—Part 2: Hygiene Requirements.* British Standard EN 1672-2: 1997, Subcommittee MCE/3/5, Food Industry Machines. English version, 1997.

European Standard, EN 1441, Medical Devices—Risk Analysis.

Food and Agriculture Organization of the United Nations (FAO). *The Use of Hazard Analysis Critical Control Point (HACCP) Principles in Food Control.* Rome: FAO, 1995.

Food and Agriculture Organization of the United Nations (FAO), Food Quality and Standards Service Food and Nutrition Division. *Food Quality and Safety Systems—A Training Manual on Food Hygiene and the Hazard Analysis and Critical Control Point (HACCP) System.* Rome: Publishing Management Group, FAO Information Division, 1998.

Gould, Brian W., Marianne Smukowski, and J. Russell Bishop. "HACCP and the Dairy Industry: An Overview of HACCP Costs and Benefits." In L. J. Unnivehr, ed. *The Economics of HACCP.* St. Paul, MN: Eagon Press, 2000.

Imholte, Thomas J., and Tammy Imholte-Tauscher. *Engineering for Food Safety.* 2nd ed. Woodinville, WA: Technical Institute for Food Safety, 1999.

International Commission for the Microbiological Specifications for Food (ICMSF). *Microorganisms in Foods 4: Application of the Hazard Analysis Critical Control Point (HACCP) System to Ensure Microbiological Safety and Quality.* Oxford, UK: Blackwell Scientific, 1988.

International Electrotechnical Commission. IEC 60601-1-1 (Ed2), *Medical Electrical Equipment, Part 1: General Requirements for Safety.* 2005.

International Organization for Standardization (ISO). *ISO 14971, Medical devices—Risk management.* Geneva: ISO, 2012.

———. *ISO 22000:2005, Food safety management systems—Requirements for any organization in the food chain.* Geneva: ISO, 2005.

———. *ISO/CD TS 22003, Food safety management systems—Requirements for bodies providing audit and certification of food safety management systems.* Geneva: ISO, 2006.

Johnson, Eric C., John H. Nelson, and Mark Johnson. "Microbiological Safety of Cheese Made from Heat-Treated Milk." *Journal of Food Protection* 53 (1990): 441–52, 519–40, 610–23.

Kraft Foods. *Supplier/Comanufacturer HACCP Standard.* Northfield, IL: Kraft Foods, February 24, 1997.

Lund, Barbara M., Tony C. Baird-Parker, and Graham W. Gould. *The Microbiological Safety and Quality of Food,* Vol. 1. Gaithersburg, MD: Aspen, 2000.

———. *The Microbiological Safety and Quality of Food,* Vol. 2. Gaithersburg, MD: Aspen, 2000.

Mihalic, Michael W., and Dennis Edwards. *Rich Products Corporation QA/Sanitation Conference Manuals.* Buffalo, NY: Rich Products Corporation, 2000.

Mortimore, Sara, and Carol Wallace. *HACCP: A Practical Approach.* 3rd ed. New York: Springer, 2013.

National Academy of Sciences. *An Evaluation of the Role of Microbiological Criteria for Foods and Food Ingredients.* Washington, DC: National Academy Press, 1985.

National Advisory Committee on Microbiological Criteria for Foods (NACMCF). *Guidelines for Application of HACCP Principles.* Washington, DC: U.S. Food and Drug Administration. August 14, 1997.

———. "Hazard Analysis and Critical Control Point Principles and Application Guidelines." *Journal of Food Protection* 61 (1998): 1246.

———. *Hazard Analysis and Critical Control Point Principles and Application Guidelines.* Washington, DC: U.S. Food and Drug Administration. August 14, 1997.

National Restaurant Association. *ServSafe Training's A Practical Approach to HACCP.* Chicago: The Educational Foundation of the National Restaurant Association, 2006.

National Seafood HACCP Alliance for Training and Education. "Record-Keeping Procedures." In *HACCP: Hazard Analysis and Critical Control Point Training Curriculum,* 1997, 111–26.

Pierson, Merle D., and Donald A. Corlett Jr. *HACCP Principles and Applications.* New York: Van Nostrand Reinhold, 1992.

Process Validation Guidance, Global Harmonization Task Force, Study Group 3, June 1, 1998.

Puri, Subhash C. *Statistical Methods for Food Quality Management.* Ottawa: Agriculture Canada Publication, 1989.

Rushing, James W., and Frederick J. Angulo. "Implementation of a HACCP in a Commercial Tomato Packinghouse: A Model for the Industry." *Dairy, Food, and Environmental Sanitation* 16, no. 9 (1996): 549–53.

Russell, J. P., ed. *The ASQ Auditing Handbook.* 4th ed. Milwaukee, WI: ASQ Quality Press, 2013.

———. *Internal Auditing Pocket Guide: Preparing, Performing, Reporting and Follow-up.* 2nd ed. Milwaukee, WI: ASQ Quality Press, 2007.

———. *The Process Auditing and Techniques Guide.* 2nd ed. Milwaukee, WI: ASQ Quality Press, 2010.

Sayle, A. J. *Management Audits.* 3rd ed. Brighton, MI: Allan Sayle Associates, 1997.

Smukowski, Marianne, and Norman Brisk. "Dairy Products Industry." In *Encyclopaedia of Occupational Health and Safety,* Vol. 3. 4th ed. Geneva: International Labor Office, 1997.

Snyder, O. Peter, Jr. *HACCP-TQM Retail Food Operations Manual.* St. Paul, MN: Hospitality Institute of Technology and Management, 1999.

Sperber, William H., et al. "The Role of Prerequisite Programs in Managing a HACCP System." *Dairy, Food, and Environmental Sanitation* 18, no. 7 (1998): 418–23.

Stevenson, Kenneth E., and Dane T. Bernard. *HACCP—A Systematic Approach to Food Safety.* 3rd ed. Washington, DC: National Food Processors Institute, 1999.

U.S. Department of Agriculture (USDA). Qualified through Verification (QTV) Program for the Fresh-Cut Produce Industry. Washington, DC: Agricultural Marketing Service, Processed Products Branch, 1999. http://www.usda.gov.

U.S. Department of Agriculture Food Safety and Inspection Service (USDA/FSIS). "Appendix A: Compliance Guidelines for Meeting the Lethality Conformance Standards for Certain Meat and Poultry Products." 1999. http://www.fsis.usda.gov/oa/fr/95033f-b.htm.

———. "Appendix B: Compliance Guidelines for Cooling Heat-Treated Meat and Poultry Products (Stabilization)." 1999. http://www.fsisusda.gov/oa/fr/95033f-b.htm.

U.S. Food and Drug Administration (FDA). *Fish and Fishery Products Hazards and Controls Guide.* 2nd ed. Washington, DC: Center for Food Safety and Applied Nutrition, 1998.

———. "Food GMP Modernization Working Group: Report Summarizing Food Recalls, 1999–2003." Washington, DC: CFSAN/Office of Scientific Analysis and Support, 2004.

———. "Foods Adulteration Involving Hard or Sharp Foreign Objects." FDA/ORA Compliance Guide, Chapter 5, Subchapter 555, Section 555.425, 1999.

———. *Guidance for Industry: Guide to Minimize Microbial Food Safety Hazards of Fresh-Cut Fruits and Vegetables.* Washington, DC: Center for Food Safety and Applied Nutrition, 2008.

———. *Guide to Inspections of Quality Systems (QSIT).* August 1999.

———. *Guidelines on General Principles of Process Validation.* May 11, 1987.

———. "HACCP Guidelines." *1999 Food Code,* Annex 5, 1999.

———. *Managing Food Safety: A HACCP Principles Guide for Operators of Food Establishments at the Retail Level.* Draft ed. Washington, DC: Center for Food Safety and Applied Nutrition, 2006.

———. "A State-of-the-Art Approach to Food Safety." *FDA Backgrounder,* August 1999.

U.S. National Archives and Records Administration. *Code of Federal Regulations.* Title 9, Parts 416 and 417. Hazard Analysis and Critical Control Point (HACCP) Systems (Meat and Poultry HACCP Regulation). 1996.

———. *Code of Federal Regulations.* Title 21, Part 110. Current Good Manufacturing Practice in Manufacturing, Packing, or Holding Food. 2000.

———. *Code of Federal Regulations.* Title 21, Part 123. Procedures for the Safe and Sanitary Processing and Importing of Fish and Fishery Products (Seafood HACCP Regulation). 1995.

———. *Code of Federal Regulations.* Title 21, Part 820. A Quality System Regulation.

Index

Note: Page numbers in *italics* indicate figures and tables.

F

G

H

I

J

K

L

M

N

O

P

Q

R

S

T

U

V

W

Y

Z